Integrated Watershed Management in Rainfed Agriculture

Integrated Watershed Management in Rainfed Agriculture

Editor

Mr. Suraj Nagar

scitus
academics

Integrated Watershed Management in Rainfed Agriculture

Edited by **Mr. Suraj Nagar**

ISBN: 978-1-68117-015-2

Library of Congress Control Number: 2015931820

SCITUS Academics LLC,
www.scitusacademics.com
Box No. 4766, 616 Corporate Way,
Suite 2, Valley Cottage,
NY 10989

Printed in United States of America on Acid Free Paper ∞

Contents

Preface

Declining groundwater is a major development concern in most areas of the world, as the livelihood of the majority of the rural population depends on sustainable water resources availability and agriculture. More productive use of rainwater is essential to mitigate the effects of water scarcity in arid regions of the planet. An integrated water management program plays an important role in building resilience in arid rural areas by improving water availability, improving food production, improving livelihoods, addressing gender issues and generating economic and social benefits for the rural population.

This book provides an extensive presentation of the improved yield of rainfed agriculture in dry and semi-arid areas. Watershed management is a constantly evolving process involving the management of other resources, biota, water and land for environmental, social and financial purposes in a defined area. In this book we examine the following questions: How has the management of water resources evolved? Which new tools are readily available and how are they really included in sustainable water supply management? To address these issues, this book covers the development of integrated strategies for sustainable management by incorporating adaptive management techniques and traditional ecological knowledge.

This book is intended for advanced graduate students, policy makers, and researchers dealing with watershed management in those areas in which farming practices rely on rainfall for water.

Chapter 1

Environmental Modeling and Exposure Assessment of Sediment-Associated Pyrethroids in an Agricultural Watershed

Yuzhou Luo[1,2], Minghua Zhang[1,2]

[1] Wenzhou Medical College, Wenzhou, China,

[2] Department of Land, Air, and Water Resources, University of California Davis, Davis, California, United States of America

ABSTRACT

Synthetic pyrethroid insecticides have generated public concerns due to their increasing use and potential effects on aquatic ecosystems. A modeling system was developed in this study for simulating the transport processes and associated sediment toxicity of pyrethroids at coupled field/watershed scales. The model was tested in the Orestimba

Creek watershed, an agriculturally intensive area in California' Central Valley. Model predictions were satisfactory when compared with measured suspended solid concentration ($R^2 = 0.536$), pyrethroid toxic unit (0.576), and cumulative mortality of *Hyalella azteca* (0.570). The results indicated that sediment toxicity in the study area was strongly related to the concentration of pyrethroids in bed sediment. Bifenthrin was identified as the dominant contributor to the sediment toxicity in recent years, accounting for 50–85% of predicted toxicity units. In addition, more than 90% of the variation on the annual maximum toxic unit of pyrethroids was attributed to precipitation and prior application of bifenthrin in the late irrigation season. As one of the first studies simulating the dynamics and spatial variability of pyrethroids in fields and instreams, the modeling results provided useful information on new policies to be considered with respect to pyrethroid regulation. This study suggested two potential measures to efficiently reduce sediment toxicity by pyrethroids in the study area: [1] limiting bifenthrin use immediately before rainfall season; and [2] implementing conservation practices to retain soil on cropland.

INTRODUCTION

Use of pesticides in crop production has been an important practice in modern agriculture, especially in the Central Valley of California, the most dynamic agricultural region in the world. Pesticide use can lead to severe environmental problems due to their toxicity to humans and many ecosystem organisms. Synthetic pyrethroids have become increasingly popular following outright bans or limitations on the use of cholinesterase-inhibiting insecticides, such as organophosphates (OPs). Previous studies have indicated that the decrease in OP use in California was related to the substitution with pyrethroids [1]. Pyrethroid insecticides are associated with selective potency in insects and relatively low potency in mammals. However, results of exposure monitoring and pesticide illness surveillance suggested that field residues of pyrethroids can cause irritant respiratory symptoms, nausea and headache [2]. Furthermore, pyrethroids are very acutely toxic to fish and invertebrates, with the 10-day LC50 values ranging from 2–140 ng/L in water (*Americamysis bahia* and *Ceriodaphnia dubia*) and 4–110 ng/g in sediment (*Hyalella azteca*) [3]. Surface water monitoring indicated widespread presence of pyrethroids and associated toxicity in agricultural and urban waterways in California [4], [5],[6]. Identifying the distribution of pyrethroids in surface waters and their effects on aquatic organisms is very important in pesticide regulation and water management of pyrethroids.

Monitoring data are usually insufficient to characterize the spatial distribution and the main sources of pesticide residues. Therefore, mathematical models are used to simulate the effects of pesticide use, management practices, and environmental factors on pesticide fate and distribution. In addition, the regulatory burden has evolved to currently consider negative impacts of pesticides on aquatic organisms. Detailed information on pesticide residues, such as the magnitude, timing and frequency of peak concentrations, are required to examine the overall ecosystem exposure by the use of pesticides. Therefore, continuous modeling at the field scale is essential for decision making processes to adequately meet regulatory requirements and improve management practices.

Recent developments in GIS technology enable the application of field-scale models on a large landscape by incorporating spatially distributed simulations of water and chemical movement in river networks. The integrated systems with field-scale models routing algorithms have been successfully applied to simulate pesticide fate and behaviors in streams. Most of those models were originally designed for the simulation of pesticides in the dissolved phase, indicating appropriate model applications on pesticides with lower adsorption coefficients. With octanol-water partition coefficients (K_{OW}) values of 10^5–10^7, pyrethroids tend to adsorb to soil and sediment rather than remain in the dissolved phase [7]. Therefore, accurate prediction of pyrethroid fate and transport must incorporate the simulations of soil erosion, in-stream sediment transport, and pyrethroid partitioning. Due to inadequate representation of the above hydrologic and transport processes, most of the existing field-scale models and in-stream routing models are not appropriate for predicting environmental behaviors of pyrethroids. For example, RZWQM (Root zone water quality model) were developed for water flow and solute transport, and thus do not simulate soil erosion and adsorbed pesticide removal. Consequently, edge-of-field pesticide fluxes are underestimated, especially for those with strong sediment/soil sorption [8]. In the GLEAMS (Groundwater loading effects of agricultural management system) model, solid-bound pesticide concentration in eroded soil is determined based on a prescribed soil mass per unit runoff volume regardless of the actual soil erosion rate [9]. In addition, many popular routing models are not able to sufficiently capture the dynamics of pesticide partitioning and transport. They either assume a steady state hydraulics (e.g., River and stream water quality model, QUAL2K [10]), or utilize prescribed suspended solid concentrations (e.g., River water quality model, RIVWQ [11]).

PRZM (Pesticide root zone model) estimates soil erosion based on a modified Universal Soil Loss Equation. In our previous study [12], PRZM model was coupled with a linear routing model for assessing pesticide dynamics and distribution in crop fields and stream networks. The coupled system provided a suitable modeling platform for determining environmental concentration and toxicity of pyrethroids. However, some model improvements were required. For example, a minor deficiency has been identified in the PRZM algorithm for adsorbed pesticide removal [13]. This paper presents an improved modeling system based on our previous study [12] for simulating the environmental fate and dispersion of pyrethroid insecticides. The specific purposes for the proposed model were to: (a) account for pyrethroid entry into surface water via soil erosion, (b) predict dynamics and distribution of pyrethroids in channel flow and bed sediment, and (c) characterize the toxicity by pyrethroids to sediment-dwelling organisms. This is one of the first studies on the dynamic modeling of pyrethroids at watershed scale, responding to the emerging research need for pyrethroid reevaluation and watershed management planning. Simulation capability of the developed model was demonstrated by applying it to the Orestimba Creek Watershed (Figure 1), an agriculturally dominated watershed in the California's Central Valley, with four pyrethroids of bifenthrin, λ -cyhalothrin, esfenvalerate, and permethrin as test agents.

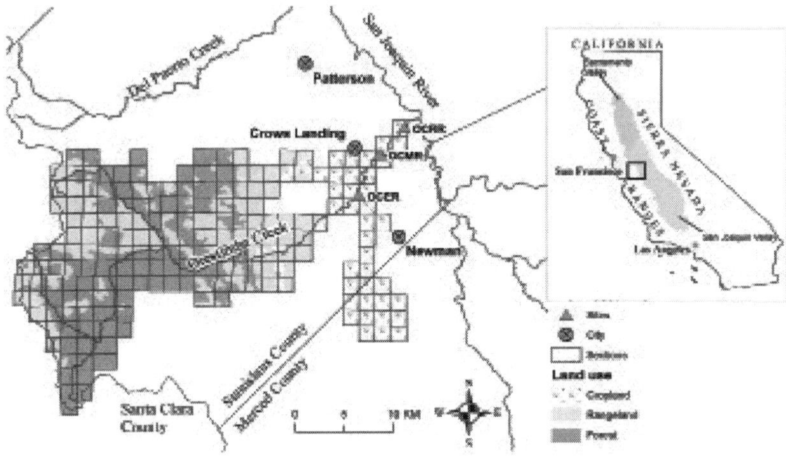

Figure 1. Orestimba Creek watershed and the sampling sites.

RESULTS AND DISCUSSION

Evaluation of the Improved Algorithm

The improved algorithm for adsorbed pesticide removal in PRZM was evaluated in a melon crop field with historical bifenthrin applications in the Orestimba Creek watershed. This field is close to the monitoring site OCER (Orestimba Creek at Eastin Road, Figure 1 and Table 1), with an average annual bifenthrin use of 0.08 kg per treated hectare. The concentration of eroded bifenthrin was reported as the total amount of eroded bifenthrin divided by the total amount of eroded soil during the simulation period. Results of the improved PRZM in this study were generally invariant with the depth of soil compartment used in the numerical analysis (Figure 2). This confirmed that the improved algorithm removed adsorbed pesticide from all compartments within the soil-interaction depth (D_E, see Materials and Methods), thus the resulting removal was not dependent with the depth of each compartment. As discussed inMaterials and Methods, the original PRZM considers only the top-most soil compartment for adsorbed pesticide removal; therefore, the results were very sensitive to the depth. The original PRZM generated similar results as the improved one when the depth of soil compartment was close to D_E (1 cm in this case study). Figure 2 demostrated PRZM simulations for soil erosion and associated pesticide removal with depth of soil compartment up to 1 cm. In the real PRZM modeling, however, small depth was required for accurate numerical simulation of water and chemical movement in the soil. For example, all crop scenarios for PRZM require depth of 0.1 cm or less for the top soil horizon developed by U.S. Environmental Protection Agency (USEPA) [14]. With small depth of soil compartment, the original PRZM significantly underestimated the adsorbed pesticide removal. The improved PRZM should be applied for consistence estimations of adsorbed pesticide release from the applied field.

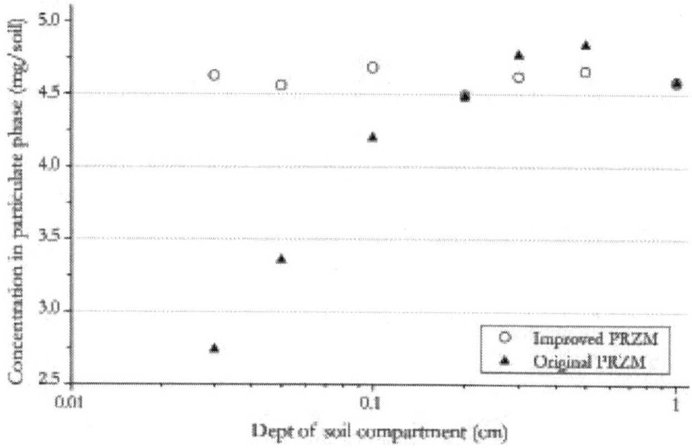

Figure 2. PRZM-simulated removals of absorbed pesticide with depths of soil compartment, tested with historical bifenthrin applications at a melon crop field in the study area.

Table 1. Sampling sites for model evaluation

Name	Site ID	USGS ID	Latitude	Longitude
Orestimba Creek at Eastin Road	OCER		37.35	—121.07
Orestimba Creek at Morris Road	OCMR		37.39	—121.04
Orestimba Creek at River Road	OCRR	11274538	37.41	—121.02

Sediment Loadings

Due to their high adsorption coefficients (K_{OW}), pyrethroids are typically adsorbed to soil particles and transported with suspended solids in surface runoff and stream flows. Therefore, a reasonable estimation of sediment concentration in a stream is the first necessary step in simulating pyrethroid partitioning between dissolved and particulate phases. Figure 3 shows the flow-weighted suspended solid concentration on a monthly basis observed and predicted at site OCRR (Orestimba Creek at River Road, close to the watershed outlet, Figure 1 and Table 1). The temporal trend of predictions followed the measured data (R^2 = 0.536), indicating a satisfactory simulation of suspended solid transport processes based on the model evaluation guidelines by Moriasi *et al.* [15]. High concentrations of suspended solids were observed during the irrigation season, especially in July.

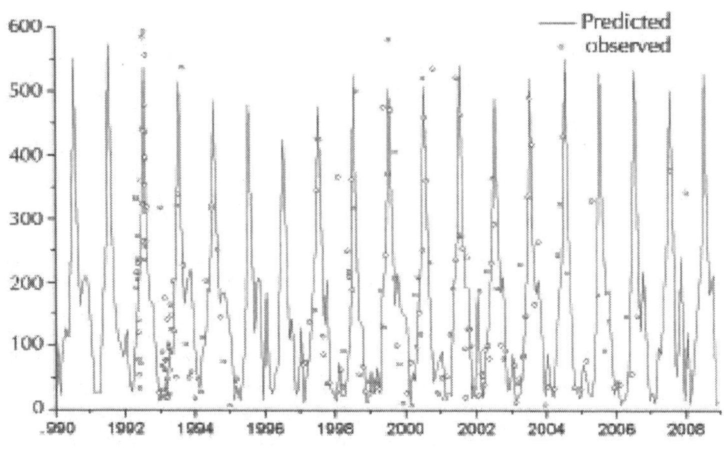

Figure 3. Observed and predicted monthly flow-weighted concentrations of suspended solids at the site OCRR (Orestimba Creek at River Road).

According to the USGS sampling results, in-stream concentrations of particulate organic carbon (OC) and suspended solids were strongly correlated (r = 0.90, p<0.001). Therefore, agreement between the predicted and observed concentrations of suspended solids also indicated a reasonable simulation for the particulate OC concentrations in the study area.

Pyrethroid Toxicity

Significant correlation ($r = 0.72$, $p = 0.004$) was observed between the predicted and measured toxicity units (TU) of pyrethroids in 15 samples at the three monitoring sites during 2007 and 2008 (Figure 4a). This indicated that the model generally captured the spatial variability and seasonality of the pyrethroid distribution in bed sediment. Based on the model prediction and field measurements, the OCRR site was generally associated with low pyrethroid TU relative to the other sites. Samples with undetected pyrethroids (plotted at 0.01 for measured TU in the figure) were all collected at the OCRR site during dormant seasons or early irrigation seasons. Located at the outlet of the Orestimba Creek, OCRR has larger drainage areas and a longer transport path for pesticides compared to other two sites. Pesticide residues have been largely decayed and diluted before reaching the water-sediment system at this site. For those undetected samples in the OCRR site, corresponding model results yielded TU values of 0.16–0.21, suggesting that the actual toxicity level was undetectable by the analytical methods applied in the sampling projects. This might also be the reason why the model overestimated measured data with low TUs. In the range of higher toxic levels (TU> = 0.3), the model had better agreement with the measurements, and the predicted and measured TUs approached the 1:1 line on the plot.

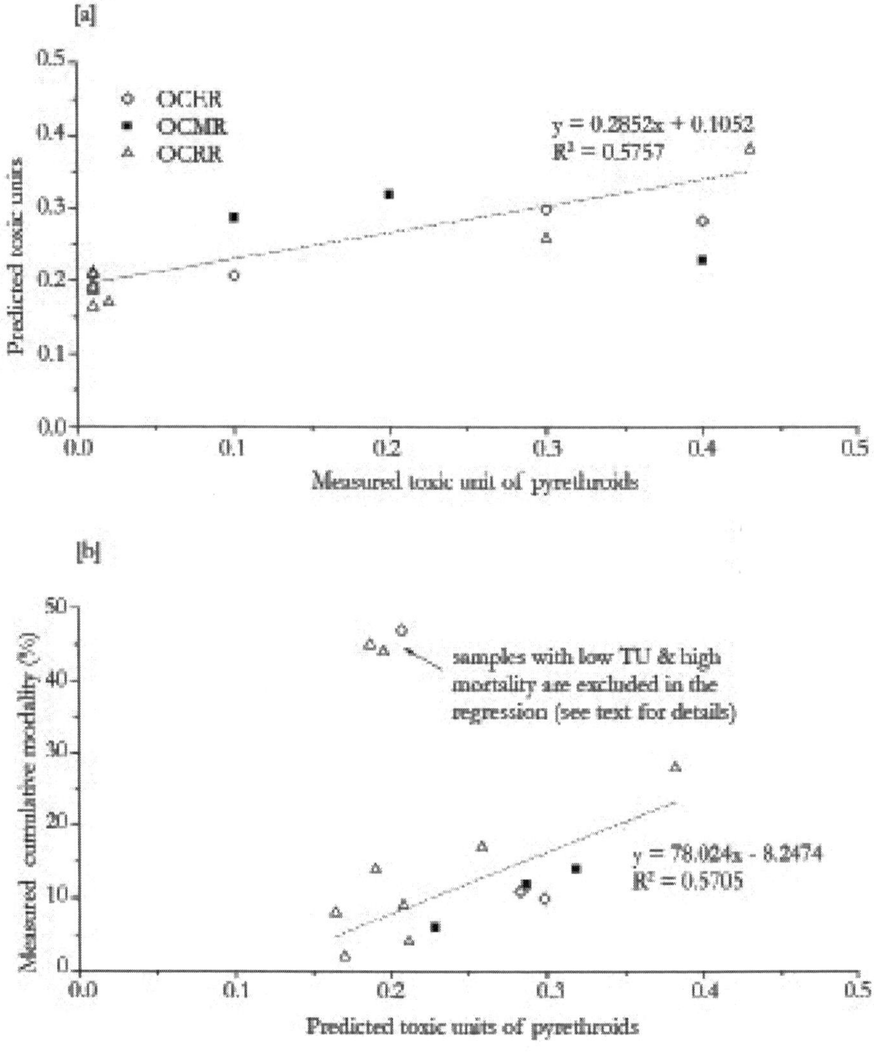

Figure 4. Predicted toxic units (TU) of the simulated pyrethroids in this study, in comparisons with [a] measured TUs (samples without detected pyrethroids are arbitrarily plotted on the figure at 0.01 TU), and [b] measured cumulative mortality.

Figure 4b compares the predicted TUs based on the simulated pyrethroids to the measured cumulative mortality of *Hyalella azteca*. For mortality values <40%, the predicted TUs are significantly correlated to the measured cumulative mortality ($r = 0.75$, $p = 0.005$). By fitting a toxicity-mortality curve in logistic form [16], the resultant

R^2 was 0.578, suggesting the model satisfactorily captured the dose-expose relationship in the evaluated sampling site. This also supported the hypothesis that sediment toxicity in the study area was mainly associated with pyrethroid concentrations. However, the correlation was not as strong among the samples with higher mortality values of >40%. For those samples, predicted TUs were about 0.2 based on the modeled pyrethroid concentrations in this study. This deficiency may have arisen due to the substantial contributions of other pesticides or other toxic compounds which were not modeled in this study to the measured sediment toxicity. This possibility was confirmed by the fact that, for those samples with high mortality, the measured TUs in sediment were also low, ranging from non-detected to 0.1. Another issue was that some samples with relatively high measured TUs were associated with low mortality. For instance, the sample at site OCMR in September 2007 had a measured TU of 0.4 and mortality of 6%. In this case, the respective model prediction of TU = 0.23 gave a more reasonable match to the observed mortality.

Characterization of Pyrethroid Exposures

In the Orestimba Creek watershed, there was a general increasing trend of total pyrethroid use during 1990–2004 (Figure 5). This increase was mainly attributed to esfenvalerate for years before 2000, and to bifenthrin and λ-cyhalothrin after 2000. After 2004, the amount of pyrethroids used has decreased, except for 2007 when reported permethrin use was very high.Figure 6 shows the predicted TUs at the OCRR site on a monthly basis, presenting sub-chronic risks of benthic organisms to pyrethroid exposures. Similar temporal trends of the predicted pyrethroid TUs were shared at the three sites in this study. Before 2000, the predicted sediment toxicity in the study was low, with maximum monthly TUs less than 0.5. Esfenvalerate was the major contributor to TUs during this period. The use of bifenthrin was started from July 1992 and explained a significant portion of the predicted TU. However, there was a general decreasing trend for both bifenthrin use (from 23.4 to 3.2 kg/year) and predicted TUs during 1992–1999. After 2000, bifenthrin use was increased again and predicted TU in sediment was substantially elevated and peaked in 2002–2004. Another potential reason for the elevated TUs was the change in application timing of bifenthrin. While it was mainly applied in July and August before

2000, significant amounts of bifenthrin were also applied during the late irrigation months of September and October and subject to the significant runoff events induced by winter precipitation. Consequently, bifenthrin became the major contributor to sediment toxicity for the last 10 years, accounting for 50–85% TU in sediment. Esfenvalerate and λ-cyhalothrin were also important contributors, especially during the irrigation season when they explained up to 50% of TU in sediment. During the recent years of 2004–2008, λ-cyhalothrin accounted for 38% of total pyrethroid used in the study area, followed by permethrin (32%), esfenvalerate (17%), and bifenthrin (10%). It is noteworthy that the use of λ-cyhalothrin was first reported in this area in 1998 and its use has been significantly increased since 2004, with an annual rate of about 180 kg in recent years. However, λ-cyhalothrin has a relatively short half-life in sediment (12 days), limiting its persistence and toxicity in aquatic ecosystems.

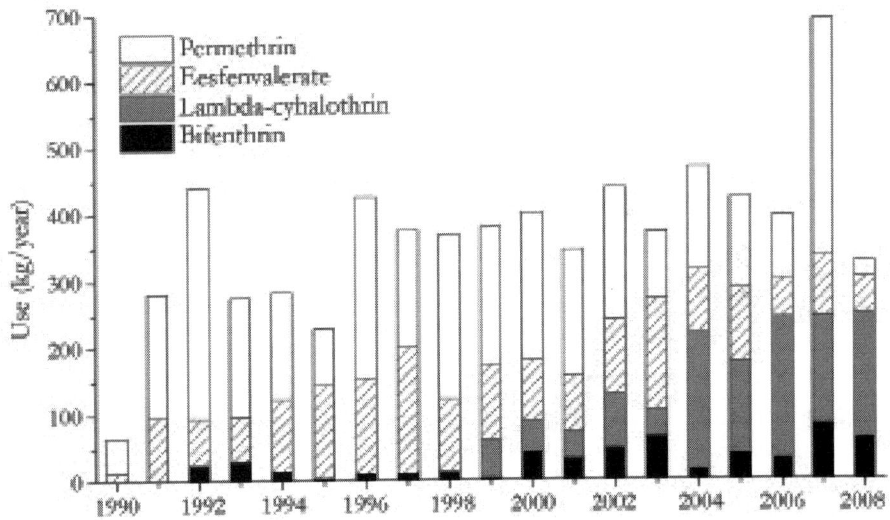

Figure 5. Annual uses of the simulated pyrethroids in the study area.

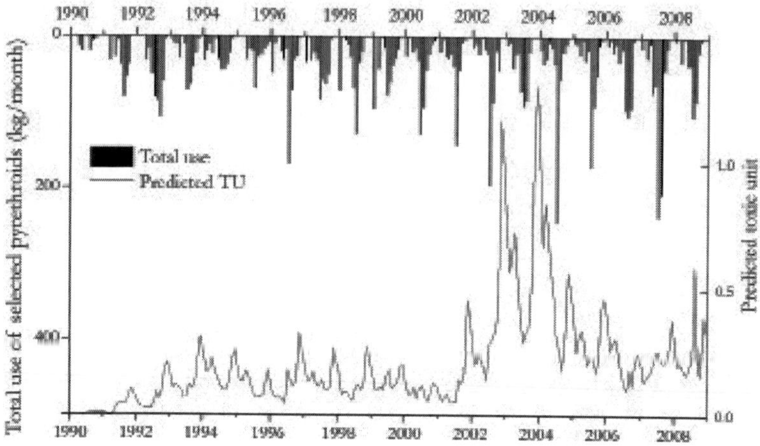

Figure 6. Predicted monthly toxic units (TU) of pyrethroids in bed sediment at the site OCRR (Orestimba Creek at River Road) and corresponding monthly pyrethroid uses in the drainage area.

During the study period, about 50% of annual pyrethroids are applied in July and August, and 70% during June to September. However, high concentrations and TU values of pyrethroids in sediment were predicted during rainfall seasons. Predicted TUs from November to January were significantly higher than the annual average. Therefore, no linear relationship was confirmed for pyrethroid uses and predicted TUs on either monthly or annual bases. In previous studies, however, such correlations were reported for organophosphate pesticides[12], [17]. With relatively high adsorption coefficients, the off-site movement of pyrethroids is mainly associated with soil erosion from agricultural fields and suspended solid transport in channels. Bifenthrin is highly persistent in soil and sediment with half-lives in aerobic soils of 85 days and in aquatic sediments of 251 days used in the modeling (Table 2). The applied pyrethroids might persist in soil and sediment long enough, and be available for the subsequent winter storms. Therefore, it is possible that predicted TUs in sediment reflected pyrethroid uses in the previous growing season. Significant time-lagged correlations were detected between the annual maximum TU and total prior bifenthrin use during the late irrigation season of September and October ($r = 0.74$, $p = 0.022$) for years when bifenthrin usage in September and October were observed (Table 3). Monthly precipitation corresponding to the maximum TU was identified as the second most important factor. Further analysis indicated that the precipitation and September+October bifenthrin use explained 90.2% of the variance of the annual maximum TUs, among which 54.9% was contributed by the bifenthrin use and 35.3% by the precipitation. This finding suggested two potential measures to efficiently reduce sediment toxicity by pyrethroids in the study area: [1] limiting bifenthrin use immediately before rainfall season; and [2] implementing conservation practices to retain soil on cropland, which would mitigate suspended soild transport to surface water bodies during early rainfall season.

Table 2. Chemical and toxic data for the simulated pyrethroids

Parameter (unit)	Bifenthrin	k-cyhalothrin	Esfenvalerate	Permethrin
Molecular weight (g/mol)	422.9	449.85	419.9	391.3
Octanol-water partition coefficient (Ko_w, Ukg)	2.00E+07	7.94E+06	1.74E+06	1.26E+06
Vapor pressure (Pa)	1.78E —05	2.00E-07	1.20E-09	2.00E-06
Henry's law constant (Pa'm³/mol)	7.74E —05	2.00E-02	4.90E —04	1.89E-01
Organic-carbon normalized partition coefficient (Ukg)	2.37E+05	1.57E+05	5300	1.00E+05
Half-life (day)				
in soil	85	25	44	42
in water	8	8	30	23
in sediment	251	12	71	40
Method detection limit (ng/g)				

in Domagalski et al. 1291	2.2	2.4	2.1	1.0
in Ensminger et al. [30]	0.5	2.0	1.0	2.0
10-day LC50 (ng/g)	5.2	4.5	15.4	108

Table 3. Precipitation, bifenthrin use, and predicted max TUs at the site OCRR during 1990–2008

Year	Max TU (rainfall season)	Precipitation (cm)		Bifenthrin use (kg)	
		Nov & Dec	**annual**	**Sep & Oct**	**annual**
1990	0.0117	1.72	17.87	0.00	0.00
1991	0.1045	2.4	23.05	0.00	0.00
1992	0.2072	2.52	22.9	0.00	24.05
1993	03128	0.89	19.16	0.00	29.41
1994	0.2629	0.71	637	0.00	13.66
1995	0.1798	3.73	29.54	1.34	2.77
1996	0.3257	12.12	41.76	0.00	10.42
1997	0.2682	12.3	26.03	0.00	9.65
1998	0.2715	3.96	47.44	0.00	10.24
1999	0.1976	2.24	28.2	0.00	3.15
2000	0.1286	0.65	19.34	8.14	41.10
2001	0.4581	9.2	26.03	0.00	31.93
2002	1.1647	11.44	24.53	13.34	48.73
2003	13027	7.23	16.9	18.73	64.19
2004	0.5627	8.49	25.67	1.34	13.52
2005	0.4622	3.92	25.67	6.38	36.93
2006	0.2462	2.8	19.01	9.60	31.01
2007	0.3816	3.47	12.91	4.02	83.24
2008	0.3923	3.26	12.95	1.76	61.13

MATERIALS AND METHODS

PRZM Application at Watershed Scale

A geo-referenced modeling system has been developed in our previous study [12] for tracking pesticide transport from its field application to the receiving waters. Pesticide discharges from the soil-canopy system were simulated by PRZM model. PRZM [18] is a one-dimensional dynamic model, primarily designed to predict the influence of climate, land/soil properties, and agricultural management on the physical and biochemical dynamics of pesticides in the environment. PRZM was selected based on its ability to simulate relevant governing processes of pesticide transport and its preferential use by the USEPA for pesticide-associated risk assessment [19]. Pre-calibrated PRZM parameters were recommended in the USEPA Standard Tier 2 scenarios for the major crops throughout the United States [14]. GIS technology was used to extend the PRZM capability for geo-referenced parameterization and application at a watershed scale.

Based on a linear routing model, edge-of-field fluxes of water and pesticides predicted by PRZM were routed through stream channels to a downstream location, e.g., a monitoring site. For water transport, stream flows at the routing destination were calculated as the summation of convolutions between PRZM-predicted runoff and corresponding watershed unit hydrograph in each simulation zone. The hydrologic response was presented by a flow-path redistribution function (*U*) based on the first passage time distribution [20], [21],

$$U_i(t) = \frac{1}{2t\sqrt{\pi(t/T_i)}}\exp\left\{-\frac{[1-(t/T_i)]^2}{4(t/T_i)\Delta_i}\right\}K_i \qquad (1)$$

where i is a running index for simulation zone, T (s) is the lag time in the flow-path, Δ_i(dimensionless) represents the shear and storage effects on the flow, and K_i (dimensionless) is the loss factor accounting for evaporation and transmission losses.

The same flow-path redistribution functions were also applied in the transport simulation of dissolved pesticides. A pesticide dispersion coefficient was determined as the sum of molecular diffusivity and flow diffusivity for the corresponding flow paths, and pesticide decay rate was calculated from its aquatic half-life. Modeling nodes in the channel network were selected to correspond with tributary/drainage junctions and monitoring locations. The developed model was applied to the Orestimba Creek watershed during 1990 through 2006, with diazinon and chlorpyrifos as test agents. The model yields reasonable agreements with measured data for the stream flow and dissolved pesticide loads [12].

Pesticide Transport with Eroded soil

PRZM is known to inadequately predict pesticide transport associated with soil erosion [13]. In PRZM, soil column was divided into compartments according to user-defined numerical simulation interval of soil depth. Adsorbed pesticide is only removed from the top-most compartment. Therefore, the removal of pesticide in adsorbed phase is primarily a function of soil compartment depth. Small depths of soil compartments, which are likely to be applied by model users to improve the numerical calculations, will result in significantly less pesticide mass removed by erosion, especially for high-sorbing compounds such as pyrethroids.

In this study, we improved the PRZM simulation algorithm by introducing a soil-interaction depth (D_E). It is assumed that all soil layers from the ground to the depth of D_E were subjected to the soil erosion process. This concept is similar to the extraction model used in transport model for estimating dissolved chemicals in surface runoff. For example, PRZM estimates the amount of dissolved pesticide runoff based on the average concentration of dissolved pesticide concentration weighted by an exponential curve for all compartments from the surface to a depth of 2 cm. The depth D_E, which could be initialized and calibrated by users, is independent from the compartment size for numerical calculation, and remains a fixed value during each PRZM simulation run.

Weighted average concentration of pesticide adsorbed on soil particles subject to erosion ($C_{S,E}$, g/g) for all compartments within the depth of D_E was first determined as:

$$C_{S,E} = \sum_{j=1}^{N_E} C_s(j).w(j) \tag{2}$$

where j is a running index for compartments, N_E is total compartments within D_E, C_s (g/g) is the concentration of soil-bound pesticides, and w is a return-to-unit weighting function, i.e., sum(w) = 1. The amount of adsorbed pesticide transported out of the field (J_{ER}, g/day) is calculated by:

$$J_{ER} = pX_e r_{om} C_{S,E} \tag{3}$$

This equation was the same as Eq. (6.13) in the PRZM manual [18], with X_e (ton/day) as the erosion sediment loss, r_{om} as the enrichment ratio for organic matter, and p as a units conversion factor (g/ton). To implement the above equation, source codes of PRZM were modified and the new procedure for determining pesticide removal with eroded soil was:

- initialize the soil-interaction depth (D_E), and define a weighting curve as a function of depth (w);
- determine the affected compartments within D_E, and calculate weighting factors for each compartment;
- adjust "B" term in the PRZM numerical solution as:

$$B\ (j)=B\ (j) + ELTERM * w(j) * DELT,\ j= 1,\ldots\ldots.N_E \tag{4}$$

where B (day^{-1}) is the diagonal element in the tri-diagonal matrix solution (Thomas algorithm) utilized by the PRZM code for the governing equations of pesticide transport, ELTERM (day^{-1}) is the erosion loss term for pesticide balance, and DELT (day) is the simulation time step.

Sediment and Pesticide Transport in Stream Network

Chemical partitioning and degradation in channel transport have been formulated in the linear routing model as described in our previous study [12]. In this study, improvements were made mainly for sediment routing and partitioning/transport of pesticide associated with suspended solids and bed sediment. A concept of sediment transport capacity was applied in this study to predict sediment deposition in the channels. By following the algorithm of Soil and water assessment tool (SWAT) [22], the maximum sediment concentration ($C_{ss,max}$, kg/m^3) that can be transported from a reach segment is calculated as:

$$C_{ss,\,max} = SPCON \times V^{SPEXP} \tag{5}$$

where V (m/s) is the peak channel velocity, and SPCON and SPEXP are coefficients to be determined. Sedimentation flux was determined by comparing the initial concentration of suspended solids in a reach at a time step ($C_{ss,0}$, kg/m^3) to $C_{ss,max}$. For instance, if $C_{ss,0} > C_{ss,max}$, the exceeding amount of suspended solids and associated pesticides in the adsorbed phase would be transported into bed sediment. The resulting sedimentation flux was used to adjust the initial concentration of suspended solids. A similar methodology was applied in the calculation of resuspension fluxes of bed sediment and pesticides by introducing factors for channel erodibility and channel cover [22]. The predicted concentrations of suspended solids were applied to determine pesticide partitioning between dissolved and adsorbed phases in the water column:

$$Fwd = \frac{1}{1 + K_d C_{ss}} \tag{6}$$

where F_{wd} (dimensionless) is the fraction of total pesticide of the water column in dissolved phase, K_d (L/kg) is the pesticide partition coefficient, and C_{ss} (kg/m^3) is the predicted concentration of suspended solids.

Pesticide simulation in bed sediment was only conducted for the active sediment layer with user-defined depth. Based on the solid-liquid partitioning, the fraction (F_{dd}, dimensionless) of total sediment pesticide in the dissolved phase was calculated as:

$$F_{dd} = \frac{1}{\phi + (1 - \phi)\rho_s K_d} \tag{7}$$

with Φ (dimensionless) denoting sediment porosity and ρ_s (g/m^3) particle density. Pesticide decay and burial in bed sediment were combined and simulated according to first-order kinetics. Pesticide transport flux by sedimentation was calculated as the product of previously determined sedimentation flux of suspended solids and the pesticide concentration in suspended solids. Similarly, pesticide resuspension flux was based on the sediment resuspension flux and the concentration of sediment-bound pesticide. Therefore, sedimentation and resuspension processes for both suspended solids and solid-bound pesticides were simulated dynamically rather than being prescribed.

Pesticide diffusion flux (J_{diff}, kg/m²/day) between the water and bed sediment was formulated using a multimedia environmental fate modeling approach [23]:

$$J_{diff} = D_{wd}\left(\frac{C_w}{Z_w} - \frac{C_d}{Z_d}\right)$$

$$D_{wd} = \left(\frac{\delta_{wd\_w}}{Z_w K_w} - \frac{\delta_{wd\_d}}{Z_d K_d}\right) \quad (8)$$

where subscripts w and d are for water compartment and sediment compartment, respectively, D_{wd} (kg/Pa/day) is the Mackay-type mass transfer coefficient, the Z's (mol/Pa/m³) are the fugacity capacity of pesticide, δ›s (m) are boundary layer depths at the water-sediment interface, and K›s (m²/day) are pesticide diffusivities. As suggested by the CalTox model [24], the boundary layer thickness in water side (δ_{wd_w}) was set as 0.02 m, while that in sediment side was estimated as $318K_d^{0.683}$.

Site Description

The modeling system newly developed in this study was applied to the field conditions of the Orestimba Creek watershed of California (Figure 1). Located in western Stanislaus County, the creek originates in the mountainous areas of the Coast Range, and discharges into the San Joaquin River. Characterized by heavier textured soils and greater slopes relative to eastside watersheds of the San Joaquin River, the Orestimba Creek watershed represents a worse-than-average condition for pesticide contamination in surface water. Climate and landscape characteristics for the studied watershed were summarized in the previous studies [12], [25].

The lower reach of the Orestimba Creek flows through agricultural lands in California's Central Valley, the most dynamic agricultural region in the world. Pyrethroids are applied to control a myriad of pests, and in this study the most important crops receiving these insecticides are orchards and row crops. Sediment from this creek was also found to be toxic to sediment-dwelling organisms, most likely because of high levels of pyrethroids [26]. Based on sediment sampling in the Orestimba Creek during 2004 irrigation season, high sediment concentrations of bifenthrin and λ-cyhalothrin were reported with acute toxicity to sensitive aquatic species[27]. In the 2010 Clean Water Act 303(d) report of California, the Orestimba Creek was listed for sediment toxicity, however the source pollutants were not yet fully identified in the report[28].

Pesticide Data Acquisition

The case study was based on two monitoring studies of pyrethroid concentrations and aquatic toxicity in streambed sediments of the Orestimba Creek watershed. Our previous monitoring study included 20 sampling sites throughout the San Joaquin River Valley, with 3 sites situated within the Orestimba Creek watershed [29] (Figure 1 and Table 1). Field measurements were conducted for 9 pyrethroids during the irrigation season of 2007. In an associated study, Ensminger et al. [30] collected monthly water and sediment samples at the site OCRR from December 2007 through June 2008, to determine concentrations of organophosphate and pyrethroid insecticides. In addition to chemical analyses, both studies conducted sediment toxicity tests with *Hyalella azteca*, following the standard USEPA protocols [31]. More details on experimental design, analytical and sediment toxicity methods, and monitoring results for the two studies can be found in Domagalski et al. [29] and Ensminger et al. [30].

Among all analyzed pyrethroids, only those detected at least twice in the two sampling studies, including bifenthrin, λ -cyhalothrin, esfenvalerate, and permethrin, were selected for model application in this study. Table 2 lists the physicochemical properties, reaction half-lives, and toxicity benchmark (as 10-d median lethal concentration, LC50, for *Hyalella azteca*in sediment) of the simulated pyrethroids. Pesticide application data were retrieved from the Pesticide Use Reporting (PUR) database maintained by California Department of Pesticide Regulation [32]. The PUR database records daily pesticide use by active ingredient and crops for each Meridian-Township-Range-Section (MTRS, or section) following the United States Land Survey System.

Simulation Design

The improved PRZM and routing simulations were performed to simulate water, sediment, and pesticide transport processes in the Orestimba Creek watershed at a daily time step for the period 1990–2008. The watershed was delineated into sections for the convenience of incorporating pesticide use data from the PUR database. Multiple fields were simulated in each section, based on the contemporary land use mapping in the study area [33]. A soil interaction depth of 1 cm and uniform weighting factors for each soil compartment were used in the case study, as suggested by SWAT documentation [22]. Individual conservation practices were not included in the model configuration. Instead, the model was calibrated based on the field measurements of water flow, sediment loading, and pesticide concentrations. Therefore, the model parameterization and simulation results reflected the overall reduction of pesticide use due to various best management practices (BMPs) implemented in the study area. Channel parameters, including Manning's roughness coefficient, flow diffusivity, coefficients for sediment transport capacity, and depth for active sediment layers, were taken from previous studies in the Orestimba Creek watershed [17], [25]. Automatic calibration was conducted for the USLE crop factor (USLE_C) to match the measured suspended solid concentrations at the watershed outlet. The calibrated model was assumed to establish a reliable hydrologic framework for the study area, and applied to the dynamic simulation of pyrethroids.

Model Evaluation

The modeling system has been validated in our previous study for its simulation capacity for stream flow and organophosphate pesticides in the dissolved phase [12]. In this study, therefore, model evaluation was emphasized on transport simulation of the suspended solids and absorbed pesticides. The location of the site OCRR is also gauged by a USGS station (#11274538) for stream flow, suspended solids, and organic carbon in suspended solids [34]. Flow-weighted concentrations of suspended solids and associated organic carbon on a monthly basis were calculated from the measurements at sampling days.

In the chemical analysis of pyrethroids, reported results were associated with different method detection limits (MDLs) (Table 2), and chemicals with concentrations lower than MDLs were reported as zeros. In addition, most of the reported concentrations of detectable pyrethroids in the study area were below the nominal reporting limit [29]. Therefore, it is not appropriate to directly compare predicted and observed concentrations for individual pyrethroids. In this study, predicted and observed pyrethroid concentrations were first converted into toxic units (TU), which is based on the assumption of toxicity additivity and is widely used as an estimate for aquatic toxicity. For each sample, the TU value was calculated as a summation of concentrations normalized by the corresponding sediment LC50 on an organic carbon (OC) basis. When no pyrethroids were detectable, the TU value was set as 0.01 for plotting convenience. To evaluate the model efficiency in predicting pyrethroid transport, the predicted TU values were compared with the measured values for each monitoring day at the three sites. It is important to note that the predicted TU values were calculated based on daily average predictions of pyrethroid concentrations, while measured values were from instantaneous samples. The model evaluation also compared the predicted TUs and cumulative mortality of*Hyalella azteca* to collected bed sediment samples. The cumulative mortality reflected the actual sediment toxicity by all chemicals, including those not analyzed or not detected, in the bed sediment. Thus, comparisons between predicted TUs and mortality were anticipated to provide useful information on the toxicity identification evaluations.

ACKNOWLEDGMENTS

The authors acknowledge the two anonymous reviewers who have helped improve the present article with their most appropriate suggestions.

REFERENCES

1. Epstein L, Bassein S (2003) Patterns of pesticide use in California and the implications for strategies for reduction of pesticides. Annual Review of Phytopathology 41: 351–375.
2. Spencer J, O'Malley M (2006) Pyrethroid Illnesses in California, 1996–2002. Reviews of Environmental Contamination and Toxicology 186: 57–72.
3. Hladik ML, Kuivila KM (2009) Assessing the Occurrence and Distribution of Pyrethroids in Water and Suspended Sediments. Journal of Agricultural and Food Chemistry 57: 9079–9085.
4. Weston DP, You J, Lydy MJ (2004) Distribution and Toxicity of Sediment-Associated Pesticides in Agriculture-Dominated Water Bodies of California's Central Valley. Environmental Science & Technology 38: 2752–2759.
5. Amweg EL, Weston DP, Ureda NM (2005) Use and toxicity of pyrethroid pesticides in the Central Valley, California, USA. Environmental Toxicology and Chemistry 24: 966–972.
6. Weston DP, Zhang M, Lydy MJ (2008) Identifying the cause and source of sediment toxicity in an agriculture-influenced creek. Environmental Toxicology and Chemistry 27: 953–962.
7. 7.Laskowski DA (2002) Physical and chemical properties of pyrethroids. Reviews of environmental contamination and toxicology 174: 49–170.
8. Ma Q, Don Wauchope R, Ma L, Rojas KW, Malone RW, et al. (2004) Test of the Root Zone Water Quality Model RZWQM for predicting runoff of atrazine, alachlor and fenamiphos species from conventional-tillage corn mesoplots. Pest Management Science 60: 267–276.
9. Leonard RA, Knisel WG, Still DA (1987) GLEAMS: Groundwater Loading Effects of Agricultural Management Systems. Transactions of the ASAE (American Society of Agricultural Engineers) 30: 1403–1418.
10. Chapra SC, Pelletier GJ, Tao H (2008) QUAL2K: A Modeling Framework for Simulating River and Stream Water Quality, Version 2.11: Documentation and User's Manual. Medford, ,

MA: Civil and Environmental Engineering Department, Tufts University.

11. Williams WM, Zdinak CE, Ritter AM, Cheplick JM, Singh P (2004) RIVWQ, Chemical transport model for riverine environments, user's manual and program documentation, version 2.02. Leesburg, , VA: Waterborne Environmental, Inc..
12. Luo Y, Zhang M (2009) A geo-referenced modeling environment for ecosystem risk assessment: organophosphate pesticides in an agriculturally dominated watershed. Journal of Environmental Quality: 38(32): 664–674.
13. USEPA (2010) Pesticide Root Zone Model (PRZM) release notes (http://www.epa.gov/ceampubl/gwater/przm3/prz3reln.html, accessed 10/2010). Athens, GA: U.S. Environmental Protection Agency, Center for Exposure Assessment Modeling.
14. USEPA (2008) USEPA Tier 2 crop scenarios for PRZM/EXAMS Shell (http://www.epa.gov/oppefed1/models/water/index.htm, accessed 09/2010). Washington, DC: U.S. Environmental Protection Agency, Office of Pesticide Programs.
15. Moriasi DN, Arnold JG, Liew MWV, Bingner RL, Harmel RD, et al. (2007) Model evaluation guidelines for systematic quantification of accuracy in watershed simulations. Transaction of the American Society of Agricultural and Biological Engineers (ASABE) 50: 885–900.
16. Scholze M, Boedeker W, Faust M, Backhaus T, Altenburger R, et al. (2001) A general best-fit method for concentration-response curves and the estimation of low-effect concentrations. Environmental Toxicology & Chemistry 20: 448–457.
17. Luo Y, Zhang X, Liu X, Ficklin D, Zhang M (2008) Dynamic modeling of organophosphate pesticide load in surface water in the northern San Joaquin Valley watershed of California. Environmental Pollution 156: 1171–1181.
18. USEPA (2006) PRZM-3, a model for predicting pesticide and nitrogen fate in the crop root and unsaturated soil zones: users manual for release 3.12.2. Center for Exposure Assessment Modeling, U.S. Environmental Protection Agency. EPA/600/R-05/111 EPA/600/R-05/111.
19. USEPA (2006) Organophosphate pesticides: revised cumulative

risk assessment (http://www.epa.gov/pesticides/cumulative/rra-op/, accessed 10/2010). Washington, DC: U.S. Environmental Protection Agency.

20. Olivera F, Maidment DR, Charbeneau RJ (1996) Spatially distributed modeling of storm runoff and non-point source pollution using Geographic Information Systems (GIS). Center for Research in Water Resources, the University of Texas at Austin (http://ceprofs.tamu.edu/folivera/UTexas/disstn/header.htm. Accessed 09/2008). CRWR Report 96-4 CRWR Report 96-4.
21. Olivera F, Maidment D (1999) Geographic information systems (GIS)-based spatially distributed model for runoff routing. Water Resources Research 35: 1155–1164.
22. Neitsch SL, Arnold JG, Kiniry JR, Williams JR (2005) Soil and Water Assessment Tool theoretical documentation, Version 2005. College Station, , TX: Agricultural Research Service and Blackland Research Centre, Texas A&M University.
23. Luo Y, Gao Q, Yang X (2007) Dynamic modeling of chemical fate and transport in multimedia environments at watershed scale–I: Theoretical considerations and model implementation. Journal of Environmental Management 83: 44–55.
24. McKone TE, Maddalena RL, Bennett DH (2003) CalTOX 4.0 a multimedia total exposure model. Berkeley, , CA: Lawrence Berkeley National Laboratory.
25. Luo Y, Zhang M (2009) Management-oriented sensitivity analysis for pesticide transport in watershed-scale water quality modeling. Water Research 157: 3370–3378.
26. Meadows R (2005) Pyrethroids in Central Valley stream sediments toxic to bottom-dwellers. California Agriculture 59: 5–6.
27. CVRWQCB (2005) Conditional Waiver for Irrigated Agriculture Monitoring Program Phase II, Quarterly Report, Activities from July 1, 2004-September 30, 2004. Sacramento, , CA: Central Valley Regional Water Quality Control Board.
28. CEPA (2010) Proposed 2010 Integrated Report (Clean Water Act Section 303(d) List/305(b) Report), http://www.waterboards.ca.gov/water_issues/programs/tmdl/integrated2010.shtml (verified 06/2010). Sacramento, CA: California Environmental Protection Agency, State Water Resources Control Board.

29. Domagalski JL, Weston DP, Zhang M, Hladik M (2010) Pyrethroid insecticide concentrations and toxicity in streambed sediments and loads in surface waters of the San Joaquin Valley, California, USA. Environmental Toxicology and Chemistry 29: 813–823.
30. Ensminger M, Bergin R, Spurlock F (2009) Pesticide Concentrations in Water and Sediment and Associated Invertebrate Toxicity in Del Puerto and Orestimba Creeks, California (http://www.cdpr.ca.gov/docs/emon/pubs/ehapreps.htm). Sacramento, CA: California Environmental Protection Agency, Department of Pesticide Regulation.
31. USEPA (2000) Methods for measuring the toxicity and bioaccumulation of sediment-associated contaminants with freshwater invertebrates (EPA/600/R-99/064). Washington DC: U.S. Environmental Protection Agency, Office of Research and Development.
32. CEPA (2010) Pesticide Use Reporting (PUR),http://www.cdpr.ca.gov/docs/pur/purmain.htm (accessed 09/2010). Sacramento, CA.: California Environmental Protection Agency, Department of Pesticide Regulation.
33. CDWR (2010) California land and water use: survey data access (http://www.landwateruse.water.ca.gov/, accessed 09/2010). Sacramento, CA.: California Department of Water Resources.
34. USGS (2010) National Water Information System: Web Interface. United States Geographical Survey, http://waterdata.usgs.gov/nwis (accessed 09/2010).
35. FOOTPRINT (2010) The FOOTPRINT Pesticide Properties Database (http://sitem.herts.ac.uk/aeru/footprint/en/index.htm, verified 05/2010). Hatfield, Herts, UK: The Agriculture & Environment Research Unit (AERU) at the University of Hertfordshire.

Chapter 2

Effects of Land Use, Topography and Socio-Economic Factors on River Water Quality in a Mountainous Watershed with Intensive Agricultural Production in East China

Jiabo Chen[1], Jun Lu[1, 2]

[1]Department of Natural Resources, College of Environment and Natural Resources, Zhejiang University, Hangzhou, Zhejiang Province, China,

[2]China Ministry of Education Key Lab of Environment Remediation and Ecological Health, Zhejiang University, Hangzhou, Zhejiang Province, China

ABSTRACT

Understanding the primary effects of anthropogenic activities and natural factors on river water quality is important in the study and efficient management of water resources. In this study, analysis of Variance (ANOVA), Principal component analysis (PCA), Pearson correlations, Multiple regression analysis (MRA) and Redundancy analysis (RDA) were applied as an integrated approach in a GIS environment to explore the temporal and spatial variations in river water quality and to estimate the influence of watershed land use, topography and socio-economic factors on river water quality based on 3 years of water quality monitoring data for the Cao-E River system. The statistical analysis revealed that TN, pH and temperature were generally higher in the rainy season, whereas BOD_5, DO and turbidity were higher in the dry season. Spatial variations in river water quality were related to numerous anthropogenic and natural factors. Urban land use was found to be the most important explanatory variable for BOD_5, COD_{Mn}, TN, DN, NH_4^+-N, NO_3^--N, DO, pH and TP. The animal husbandry output per capita was an important predictor of TP and turbidity, and the gross domestic product per capita largely determined spatial variations in EC. The remaining unexplained variance was related to other factors, such as topography. Our results suggested that pollution control of animal waste discharge in rural settlements, agricultural runoff in cropland, industrial production pollution and domestic pollution in urban and industrial areas were important within the Cao-E River basin. Moreover, the percentage of the total overall river water quality variance explained by an individual variable and/or all environmental variables (according to RDA) can assist in quantitatively identifying the primary factors that control pollution at the watershed scale.

INTRODUCTION

The deterioration of river water quality has become a primary environmental concern due to unsustainable anthropogenic activities; the demand for freshwater has been rapidly increasing in many developing countries, especially in China [1]–[3]. River water quality is controlled by complex anthropogenic activities and natural factors at both the river and watershed scales [1],[7]–[9]. Understanding the temporal and spatial variations in river water quality and estimating the primary regional factors that affect water quality can assist researchers in establishing priorities for sustainable water management [2], [10]–[12]. The relationships between water quality parameters and land use/cover, population density and point source discharge have been frequently studied [1], [7], [9], [13]–[15]. Simultaneously, topography and animal waste discharge are considered important factors that affect watershed river water quality [7],[13]–[16]. Particularly, several countries, e.g., China, contain large mountainous areas [17] and extensive animal production that lack strict management techniques [16], [18], [19].

Recently, Pearson correlations have been widely employed to determine the relationships between environmental variables and river water quality [1], [7]. This method is simple and provides quantitative information; however, Pearson correlations lack visualization. Multiple regression analysis (MRA) is a useful tool that is commonly used to determine that watershed characteristics that best explain the spatial variability of an individual river water quality variable [7], [20]. This method lacks general information regarding pollution types. Moreover, principal component analysis (PCA) has become a widely accepted method in river water quality assessment and source apportionment studies in the last decade [2]. This method is commonly used to obtain the types of specific pollution sources without explanatory variables; however, PCA can only provide preliminary information regarding pollution types. Redundancy analysis (RDA), a multivariate statistical analysis method, has been proved useful for qualitative analysis of the interactions between river water quality and watershed characteristics in highly complex systems [13], [21], [22]. RDA can reveal the influences of environmental factors on overall water quality, not only on single water quality variables [20]. Several commonly used statistical methods for pollution source identification are complementary (i.e., they have their own benefits and limits). Summary of commonly used statistical methods on pollution source identification in recent years is listed in **Table S1**. However, comprehensive applications of different statistical methods to evaluate the effects of environmental variables on water quality have not been fully explored in river studies in China. Furthermore, few studies have used RDA analysis to quantitatively evaluate the effects of watershed characteristics on the overall river water quality.

The Cao-E River is located upstream of the Qiantang Estuary, which is a major riverine system in Zhejiang Province, China, and is the primary source of industrial, agricultural and domestic water supplies [10]. Furthermore, the blue algae bloom that occurred in July 2004 occurred in the Qiantang Estuary [23], [24].

In this study, ANOVA, Pearson correlation analysis, MRA, PCA and RDA were applied to investigate the effects of sub-watershed land use, topography and socio–economic factors (including animal waste discharge) on river water quality at the watershed scale for the Cao-E River system in a GIS environment. The objectives of this study were threefold: (1) to examine the temporal and spatial variations in water contamination within the study area, (2) to investigate the relationships between the river water quality parameters and land use, topography and socio-economic factors, and (3) to identify the primary pollution sources to estimate the possible sources that affect the water quality parameters in the Cao-E River basin. The results can be helpful for water conservation in the Cao-E River basin, and they provide a valuable tool for water quality agencies to develop assessment strategies for effective water quality management and rapid solutions for water pollution problems at the watershed scale.

MATERIALS AND METHODS

Ethics Statement

No specific permits were required for the described field studies; the sampling did not cause any disturbance to the environment or to the protected species at the sampling sites.

Study Area

The Cao-E River (29°08′-30°15′ N, 120°30′-121°15′ E) is located in Zhejiang Province, East China. The Cao-E River is one of the main tributaries of the Qiantang Estuary, which flows into the East China Sea. The river system contains a main stream and several major tributaries (i.e., the CT River, the XC River, the CL River, the HZ River, the XS River and the YT River;**Figure 1**). The main stream is ~197 km long with an average slope of 3.0‰. The Cao-E River basin has a typical subtropical monsoon climate with an average annual rainfall of 1500 mm and an average ambient temperature of 16.2°C [25]. The watershed is within a mountainous region. Specifically, mountains and hills cover 2/3 of the area; the remaining area is covered by plains with intensive agricultural production [26].

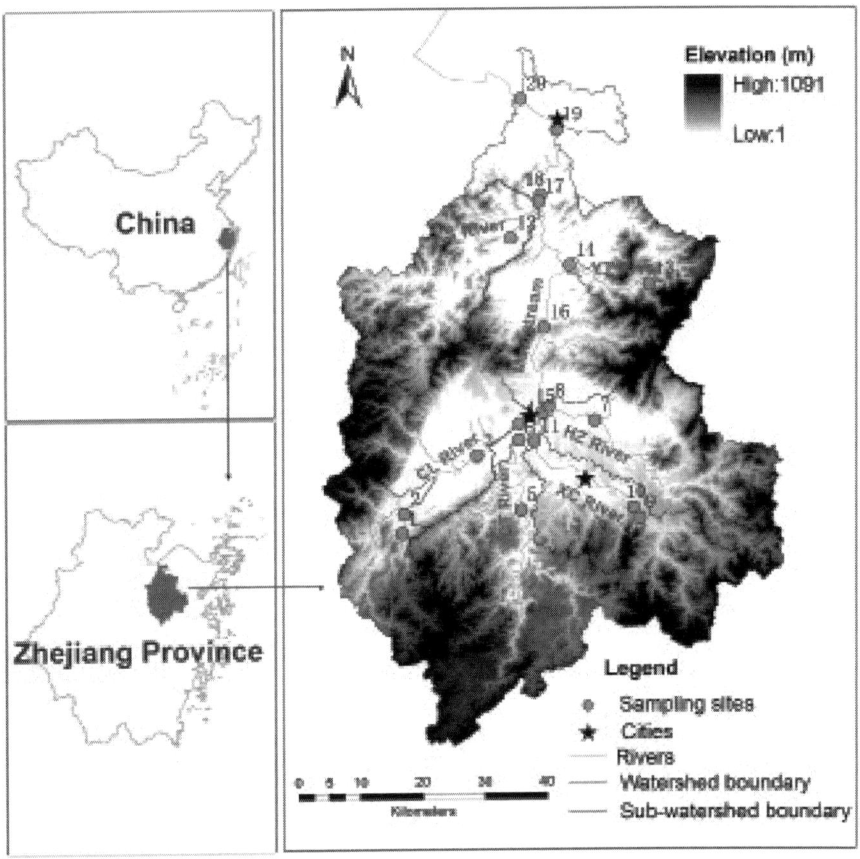

Figure 1. The Geographical Location Of The Sampling Sites In The Cao-E River System. (Sampling Stations 1-Cl1, 2- Cl2, 3-Cl3, 4-Cl4, 5- Ct 1,6-Ct2,7-Hz1,8- Hz2,9- Xc1,10- Xc2,11- Xc3,12- Xs1,13- Yt1,14- Yt2,15- M1,16-M2,17- M3,18- M4,19- M5, 20- M6).

Data Sources

Water Quality data

The stream water samples (the number of samples used for statistical analysis is 36) were collected from July 2003 to June 2006 at 20 sampling sites throughout the watershed (Figure 1). Specific sampling stations that were largely influenced by point source pollution were excluded from further analysis. These specific sampling sites should meet the following conditions: (1) there was an obvious outlet of point pollution discharge near the sampling site, and (2) The standard deviation of the water quality parameter (BOD_5) at the specific sampling sites were triple more than the standard deviation for total sampling sites. The samples were collected once a month between 9:00 am and 16:00 pm. Moreover, the water samples were collected at approximately 0.3 m below the water surface in the central stream, placed into plastic bottles (2.5 L), transported to the laboratory and stored at 0~4°C for subsequent chemical analysis. The chemical measurements were performed in the laboratory within 24 h after collecting the water samples. Moreover, dissolved oxygen (DO), water temperature (T), pH and electrical conductivity (EC) were measured when the water samples were collected with a hand-held multi-parameter instrument (Multi 340i SETs, The Merck Co. Ltd., Germany). The turbidity was measured with a hand-held Turbiquant 1000IR (Multi 340i, The Merck Co. Ltd., Germany). The following chemical and biological water quality parameters were measured: chemical oxygen demand (COD_{Mn}), 5-day biochemical oxygen demand (BOD_5), total nitrogen (TN), total phosphorus (TP) and dissolved phosphorus (DP), which were measured using standard methods [27]. The ammonium nitrogen (NH_4^+-N) content was measured using an Astoria analyzer system (AAS, Brown Rupee CO. Ltd., Germany) after filtration through a 0.45-μm filter (Hailing Medicine, Zhejiang Province, China). The nitrate nitrogen (NO_3^--N) content was measured from the absorbance of the sample at 220 nm (A220) and 275 nm (A275) using a visible and ultraviolet spectrophotometer (Shanghai Spectrum Instruments, Shanghai, China) after filtration through a 0.45-μm filter; the results were determined by the following relationship: A220 - 2×A275 [27].

Land use/cover, Topography and socio-Economic data and GIS Analysis

Latitude and longitude were measured using a hand-held GPS. The catchment boundaries were delineated on a 30-m spatial resolution digital elevation model (DEM); the stream network was represented using the Soil and Water Assessment Tool (SWAT) model. ArcGIS 9.3 Desktop GIS software was used to calculate the watershed characteristics for each sampling site. All datasets were converted to a common digital format (WGS 1984) and a common coordinate system (Albers Equal Area). Hydrologic units were not solely defined by the watershed above an individual point [28]; a hydrologic unit on the Cao-E River may cover 2/3 or more of the entire watershed above that sampling station (most hydrologic units on tributaries encompass the entire watershed above a sampling station).

The socioeconomic data, including land use change, the gross domestic product, animal husbandry output and human population in the watershed, were obtained from the Shengzhou (SZ), Xinchang (XC), Shangyu (SY) and Shaoxing (SX) County Statistical Yearbooks for the period 2003–2006 and were adjusted with some typical investigation in several towns within the watershed. The temporal variations in land use were considered to be minimal. The annual changes in forest areas and water areas were less than 0.05%, the annual decrease rate in cropland areas was less than 1% and the annual increase rate in urban areas was less than 2.5%., respectively, for the entire watershed and throughout the study period. Thus, Landsat Thematic Mapper imagery from 2006 was used to map the land cover in the study area [29]. The land cover was categorized into the following 4 classes: forest, water, cropland and urban land. The annual increases in human population density, gross domestic product per capita and animal husbandry output per capita were less than 1.5%, 10% and 2%, respectively, for the entire watershed and throughout the study period. Therefore, the averages of the socio-economic data for the entire study period were selected in this study. The slope map was derived from the DEM using a GIS spatial analyst tool (ESRI, 2006). The land use, slope, DEM and administrative map (with attribute data, including the gross domestic product, animal husbandry output and human population for every town) were used to calculate the respective land use area (forest, water, urban land and cropland), mean elevation, mean slope, area, human population density, gross domestic product per capita and animal husbandry output per capita within each sub-watershed using a GIS spatial analyst tool. The abbreviations and statistics of the sub-watershed characteristics in the Cao-E River basin are summarized in**Table 1**.

Table 1. Abbreviations And Basic Statistical Information Of The Sub-Watershed Characteristics In The Cao-E River Basin.

	Parameters	Abbreviation	Range	Mean -± 5d
Land Use	Forest Percentage (96)	En	25.72-74.50	50.31=12.36
	Water Percentage (%)	Wt	0.13-3.39	1.87⁴ 0.87
	Urban Percentage HO	Ub	0.45-14.44	4.09=352
	Cropland Percentage (%)	Crp	21.58-57.73	43.85= 10.18
Topography	Mean Elevation (M)	Elv	125.12-55654	306.171·116.17
	Mean Catchment Slope ( )	Snp_{mean}	10.20-17.72	13.74=2.29
	Catchment Area (Km^2)	Ar	3839-228621	76453·1674.58
Socio-Economic Factors	Human Population Density (Persons/ Km^2)	Hp_d	123.01-576.68	316.111119.12
	Gross Domestic Product Per Capita (10^5 Chinese Yuan)	Gdp_{pc}	13551-2151.16	58238=427.88

	Animal Husbandry Output Per Capita (10^5 Chinese Yuan)	Aho_{pc}	8.07-39.40	205618.61

Data Analysis

Analysis of variance (ANOVA) was used to compare river water quality variations between different seasons (i.e., rainy and dry) via the Student-Newman-Keuls multiple-range test (P<0.05). Relationships among the considered variables were tested using Pearson's coefficient with statistical significance set at P<0.05. Major gradients and principal patterns in the water quality data between wet and dry seasons were detected using principal component analysis (PCA). PCA transforms a dataset consisting of *p* variables (analytical constituents) that are interrelated or correlated to various degrees into a new dataset containing *p* new orthogonal and uncorrelated variables; these variables are called the principal components (PCs) [30], [35]. The PCs are linear functions of the original variables in which the sum of their variances is equal to that of the original variables [30]. The PCs following a descending ordered according to their variances with PC1 corresponding to the variable with the largest variance. Algebraically, for *p* original variables (i.e., $x_1, x_2, \ldots, x_p$). Additional PCs up to *p* can be formulated in a similar manner. The variances of the PCs are the eigenvalues; the coefficients or weights, are the eigenvectors extracted from the covariance or correlation matrix. In the example PCA, the correlation matrix was used in all cases. The goal of PCA is that the first *k* PCs (where $k<<p$) retain most of the information in all of the *p* original variables, which effectively reduces the practical dimensionality of the dataset. More specifically, if the correlations are high among many of the original variables, the first few PCs contain (or explain) a large percentage of the total variance and may be used to describe multivariate patterns or water quality variations across the watershed nearly as well as using the complete set of *p* original variables. These patterns are often related to specific sources of contamination [30].

The stepwise regression analysis was performed to determine the environmental or socio-economic factors (TP, P<0.10 and the others, P<0.05) that best explained the spatial variability of an individual river water quality variable [20]. Variables that were strongly intercorrelated with others (variance inflation factor >10) in the initial analysis, were removed and a further analysis was carried out with the remaining environmental variables for MRA and RDA [20], [31], [32]. Variance inflation factor (VIF) is a common way for detecting multicollinearity, a general rule is that the Variance Inflation Factor should not exceed 10 [49]. RDA was performed using Canoco 4.5 software to evaluate the relationships between river water quality and the environmental variables [33]. RDA was chosen because previous inspection of the data revealed a linear response rather than a unimodal response in the primary river water quality variables [34]. A Monte Carlo permutation test was used to verify the significance of the models [29]. Two series of matrices (i.e., for river water quality and environmental variables) from the RDA results were visualized as ordination diagrams using the CanoDraw software package for Windows. The river water quality parameters and the environmental/socio-economic variables were represented with arrows. Correlations between the watershed characteristics and/or the river water quality parameters were obtained by projecting the watershed characteristics onto each river water quality parameter in which higher values indicated higher correlations [34]. Application summary of 5 statistical methods in this study was listed in **Table S2**. The statistical analyses, except RDA, were completed using the SAS 9.1 software for Windows.

RESULTS

Physico-Chemical water Quality in the Cao-E River basin

ANOVA was used to compare river water quality variations between the different seasons (i.e., rainy and dry). Temporal and spatial variations in the physico-chemical parameters are shown in **Table 2** and **Figure 2**. More specifically, BOD_5, TN, DO, pH, T and turbidity exhibited significant temporal variations (**Table 2**, $p<0.10$). Moreover, TN, pH and T were generally higher in the rainy season (April to September), whereas higher values for BOD_5, DO and turbidity occurred in the dry season (October to March). Most of the water quality parameters, except pH and T, exhibited considerable spatial variations. BOD_5, COD_{Mn} and NH_4^+-N were higher in urban areas compared with surrounding rural areas, whereas DO exhibited a reverse response to the extent of urbanization. TN, DN, NO_3^--N, TP and DP were higher in the lower part of the basin where more land has been developed and were lower in the upper-eastern part of the Cao-E River basin. Turbidity and EC were higher in the lower part of the basin (**Figure 2**) where the reach is often affected by tides and/or point source pollution [36], [37].

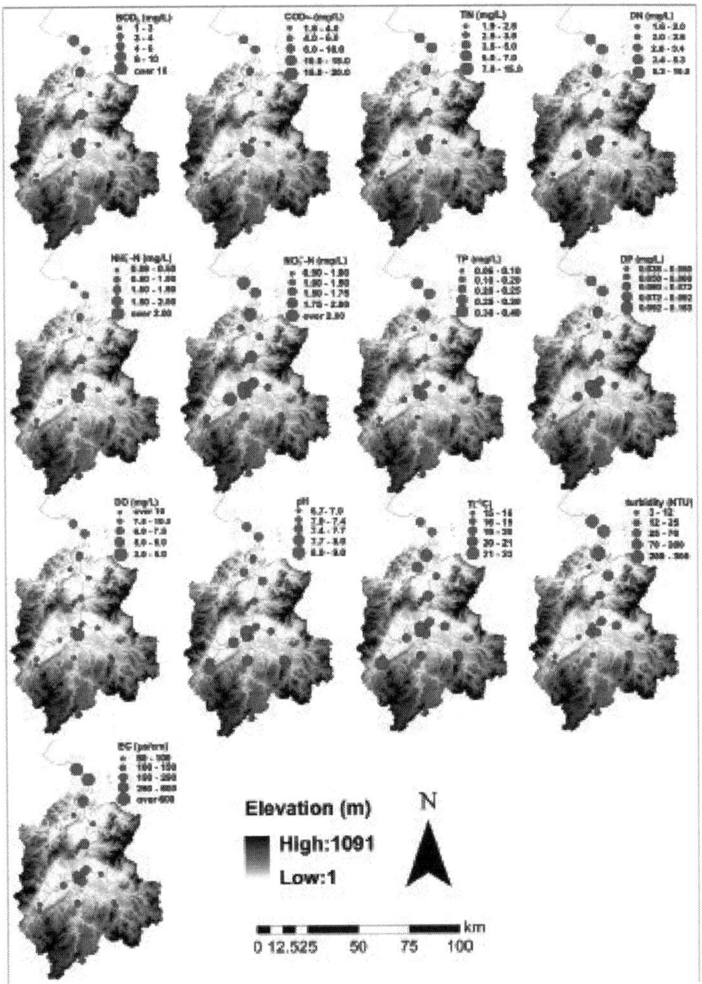

Figure 2. Spatial Variations In Bod5, Codmn, Tn, Dn, Nh_4^+-N, No_3^--N, Tp, Dp, Do, Ph, T, Turbidity And Ec In The Study Area.

Table 2. Seasonal Averages, Coefficients Of Variation [Cv (%)] And River Water Quality Parameter Ranges

Water Quality Parameters	Rainy Season		Dry Season		
	Average [Cy (%)]	Range	**Average [Cy (%)]**	**Rang.**	***P-Value***
$Bori_s$ (Mg/L)	5.912 [183]	0.05-173.5	7.850 [181]	0-255.00	0.052
Cod_{mr}, (Mg/L)	3.845 [87]	039-43.87	4.0361 [99]	0.49-47.75	0.557
Tn (Mg/L)	4.793 1103]	0.29-45.85	4.208 [73]	0.56-31.09	0.049
Dn (Mg/L)	3549 [90]	0.06-22.53	3.295 [80]	0.03-30.07	0.219
Nh_4^+ -N (Mg/L)	0.605 [195]	0-15.87	0.759 [2099)	0-19.58	0.156
No_3^--N (Mg/L)	1.677 [41]	0.01-4.30	1.606 [69)	0.08-15.08	0.283
Tp (Mg/L)	0.150 [108]	0-1.88	0.168 [113]	0-3.86	0.317
Dp (Mg/L)	0.061 (911	0-0.49	0.076 [253]	0-3.49	0.163
Do (Mg/L)	7.164 [33]	0-19.55	8.045 [42]	0-17.86	–0.001
Ph	7.538 (7]	533-9.34	7.449 [8]	5.45-10.12	0.055
T (˜C)	24.967 (24)	8.40-37.20	14.262 [64]	4.70-105.00	–0.001
Turbidity (Ntu)	70.011 [105]	0.83-586.10	92.939 [113]	0.91-1111	0.002
Ec (μs/Cm)	306.38 [347]	14-14420	348.56 [282]	14-11230	0.557

Major gradients and principal patterns in the water quality data between the wet and dry seasons were detected using PCA. The PCA based on only the dry season water quality data indicated that 4 significant factors (i.e., PCs with an eigenvalue >1) explained 87.8% of the total variance (**Table 3**). The first factor accounted for 50.2% of the total variance and had a strong positive correlation with BOD_5, COD_{Mn}, TN, DN, NH_4^+-N, and T. The PCA of the wet season water quality data showed that the eigenvalues for the 2 most significant factors were 8.442 and 1.817, accounting for 78.9% of the total variance. Factor 1 accounted for 64.9% of the total variance and was positively correlated with BOD_5, COD_{Mn}, TN, DN, TP and DP

Table 3. River Water Quality Variables (13) On Rotated Pcs For Datasets During The Dry And Wet Seasons

Water Quality Parameters	Dry Season				Wet Season	
	Pc1	Pc2	Pc3	Pc4	Pc1	Pa
Bod_s (Mg/L)	0.354	-0.110	0.008	0.076	0.312	-0.041
Cod_{mn} (Mg/L)	0.350	0.200	-0.187	0.128	0310	0.263
Tn (Mg/L)	0.380	0.001	0,077	0.083	0.333	-0.054
Dn **(Mg/L)**	0.362	-0.044	0.147	0.127	0311	-0.197
Nh_4^+**-N (Mg/L)**	0.345	-0.215	0.019	0.084	0.267	-0.346
No_3 -N (Mg/L)	0.275	-0.020	0.425	-0.079	0.283	0.174
Tp (Mg/L)	-0.221	0.409	0.077	-0.476	0.314	0.200
Dp (Mg/L)	0.097	0.329	0.609	-0.090	0.300	-0.018
Do (Mg/L)	-0.270	0.361	0.091	-0.204	0.272	0349
Ph	-0.146	0.328	0.268	0.614	-0.184	0.546
T (°C)	0.418	-0.420	0.119	-0.212	0.010	0.246
Turbidity (Ntu)	0.222	0.340	-0.416	-0.312	0.283	0.246
Ec (ms/Cm)	0.241	0.308	-0.335	0.382	0.247	0395
Eigenvalues	6.522	2.344	1.518	1.028	8.442	1.817
Percentage Of Variance	0.502	0.180	0.117	0.079	0.649	0.140
Cumulative Percentage **Of Variance**	0502	0.682	0.799	0.878	0.649	0.789

Pcs Refer To Principal Components. The Significant Factors (I.E., Pcs With An Eigenvalue .--1) Are List In The Table.

Relationships between Watershed Characteristics and river water Quality

Relationships between the river water quality variables and their corresponding sub-watershed land use, topography and socio-economic factors were explored using a Pearson correlation test, MRA and RDA. All environmental variables, including land use, topography and socio-economic factors, were utilized to evaluate the response of river water quality to environmental gradients. The correlation analysis indicated that CRP was positively correlated with TN, DN, NO_3^--N, TP and DP (r>0.45, p<0.05; r refers to the correlation coefficient). Moreover, UB was positively correlated with BOD_5, COD_{Mn}, TN, DN, NH_4^+-N, TP, DP and EC (r>0.50, p<0.05) and negatively correlated with DO and pH (r<−0.65, p<0.01). FRT was positively correlated with DO and pH (r>0.47, p<0.05) and exhibited a negative correlation with BOD, COD, TN, DN, NH_4^+-N, NO_3^--N TP, and DP (r<−0.50, p<0.05). Furthermore, SLP_{mean} was positively correlated with DO (r = 0.59, p<0.01) and negatively correlated with COD_{Mn}, TN, DN, NH_4^+-N, NO_3^--N, TP, DP, T, turbidity and EC (r<−0.48, p<0.05). The elevation was positively correlated with DO (r = 0.45, p<0.05) and negatively correlated with COD_{Mn}, TN, DN, NO_3^--N, TP, DP, turbidity and EC (r<−0.47, p<0.05). Additionally, GDP_{pc} exhibited a positive correlation with COD, TN, DN, NO_3^--N, TP, DP, turbidity, and EC (r>0.44, p<0.05). AHO_{pc} was positively correlated with TN, NO_3^--N, TP, DP, turbidity and EC (r>0.47, p<0.05). Lastly, HP_d exhibited a negative correlation with DO and pH (r<−0.52, p<0.05) and a positive correlation to all other river water quality parameters (r>0.44, p<0.05) (**Table 4**).

Table 4. The Pearson Correlation Coefficients For The River Water Quality Parameters Compared With The Corresponding Sub-Watershed Characteristics

	Frt	**Wt**	**Ub**	**Crp**	**Elv**	**Slp_{mean}**	**Ar**	**HP_d**	**Gdp_{pc}**	**Aho_{pc}**
Bod_5	-0.52^{*}	0.11	0.76^{**}	035	-0.31	-0.40	-0.08	0.59^{**}	0.32	0.09
Cod_{mn}	0.61^{**}	0.31	0.90^{**}	0.40	0.58	-0.64^{**}	0.10	0.78^{**}	0.60^{**}	0.42
Tn	-0.72^{**}	0.27	0.96^{**}	051^{*}	-0.61^{**}	-0.74^{**}	0.37	0.87^{**}	0.60^{**}	0.48^{*}
On	-0.70^{**}	0.17	0.91^{**}	0.51^{*}	-0.53^{*}	-0.64^{**}	0.34	0.81^{**}	0.48^{*}	0.40
Nh_4^+-N	-0.60^{**}	0.15	0.84^{**}	0.42	-0.39	-0.51^{*}	0.14	0.69^{**}	0.38	0.19
Nh_3^--N	-0.54^{*}	0.13	0.41	0.50^{*}	-0.52^{*}	-0.76^{**}	0.70^{**}	0.60^{**}	0.45^{*}	0.70^{**}
Tp	-0.63^{**}	037	0.77^{**}	0.46^{*}	-0.82^{**}	-0.76^{**}	0.46^{**}	0.77^{**}	0.66^{**}	0.80^{**}
Dp	-0.60^{**}	0.08	0.61	0.50^{*}	-0.48^{*}	-0.63^{**}	0.30^{*}	0.64^{**}	0.45^{4}	048^{**}
Do	0.48^{*}	-043	-0.75^{**}	-0.29	0.45^{*}	0.59^{**}	-0.63^{**}	-0.71^{**}	-0.44	-0.36
Ph	0.53^{*}	0.01	-0.66^{**}	-0.42	0.37	0.33	0.44	-0.53^{*}	-0.16	0.15
T	-0.43	0.18	0.44	0.36	-0.44	-0.49^{**}	0.46	0.53^{*}	0.23	0.33
Turbidity	-0.28	0.40	0.42	0.15	-0.71^{**}	-051^{*}	0.36	0.45^{*}	0.58^{**}	0.81^{**}
Ec	-0.21	0.61^{**}	0.52^{**}	0.02	-0.64^{**}	-0.60^{**}	0.20	057^{**}	0.88^{**}	0.78^{**}

Abbreviations Are Provided In Table 1. The Figures In The Table Are Correlation Coefficients.

*Indicates Significance At The 0.05 Probability Level.

**Indicates Significance At The 0.01 Probability Level.

MRA demonstrated that no individual environmental factor was able to describe the overall water quality; however, most of the physico-chemical water parameters could be sufficiently predicted using 1 to 3 environmental factors (**Table 5**). Specifically, BOD_5 could be predicted using UB, AR and GDP_{pc}; COD_{Mn} with UB and AR; TN, DN, pH and DO with UB; NH_4^+-N with UB and HP_d; DP with HP_d; turbidity with AHO_{pc}; EC with GDP_{pc}; TP with UB and AHO_{pc}; and NO_3^--N with UB and SLP_{mean}.

Table 5. Stepwise Multiple Regression Models For The River Water Quality Parameters And Watershed Characteristics In The Cao-E River Basin In Eastern China

Parameters	Independent Variables	Regression Equation	R^2	Adjusted R^2	Significance
Bod_5	Ub, Ar, Gdp_{pc}	$1.053+566.924ub-0.012ar-0.015gdp_{pc}$	0.856	0.830	**
COdmn	Ub, Ar	1.451+88.727ub-0.002ar	0.903	0.892	**
Tn	Ub	1.668+69.168ub	0.912	0.907	**
Dn	Ub	1.231+53.629ub	0.821	0.811	**
Nh_4^+-N	Ub, Hp_d	$1.407+61.854ub-1.030hP_d$	0.805	0.782	**
No_3^--N	Slp_{mean}, Ub	$5.161012-0.2335slp_{mean}-7.675ub$	0.720	0.670	**
Tp	Ub, Aho_{pc}	$0.012+1.093ub+0.050aho_{pc}$	0.794	0.770	*
Dp	Hp_d	$0.022+0.015hp_d$	0.406	0.373	**
Do	Ub	9.802-50.534ub	0.562	0.538	**4
Ph	Ub	7.724012-5.814ub	0.439	0.408	**
Turbidity	Aho_{pc}	$—115.264+96.209aho_{pc}$	0.662	0.643	**
Ec	Gdp_{pc}	$—261.761+1.009gdp_{pc}$	0.778	0.766	**

Abbreviations Are Provided In Table 1. The Parameters Without Regression Models Are Not Listed.

*Indicates Significance At The 0.05 Probability Level.

**Indicates Significance At The 0.01 Probability Level.

The RDA of the overall water quality as the dependent variable suggested that FRT, UB, CRP, ELV, SLP_{mean}, AR, HP_d, GDP_{pc} and AHO_{pc} explained 15.7–50.8% of the river water quality spatial variation, and a combination of the topography, land use and socio-economic factors explained 86.1% of the variation in overall water quality (**Table 6**). The ordination diagram of overall water quality and topography, land use and socio-economic factors indicated that the first RDA axis displayed a pollution gradient (e.g., COD_{Mn} and TN increased along the axis), which was positively correlated with UB and HP_d and negatively correlated with FRT and SLP_{mean} (**Figure 3**), accounting for 52.4% of the total variance in water quality (**Table 7**). The second axis was related to AHO_{pc} and only explained 16.5% of the total variance (**Table 7**).

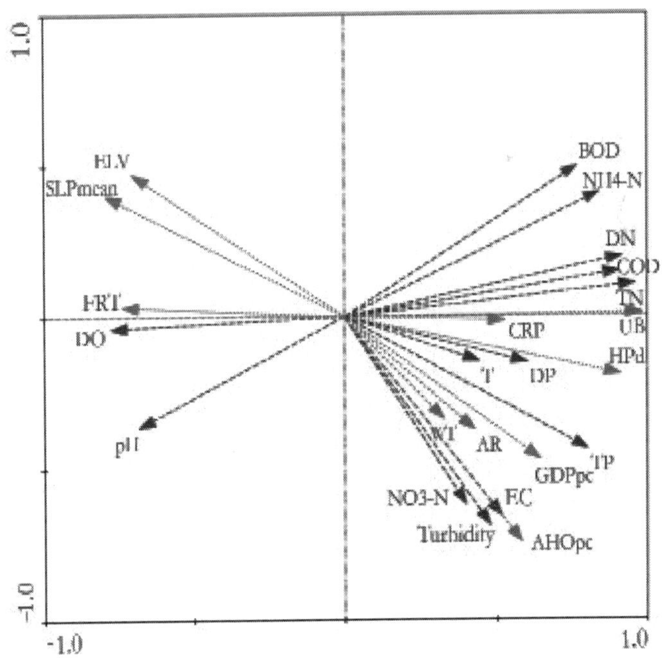

Figure 3. Biplots Of The River Water Quality Parameters And Watershed Characteristics In The Cao-E River Basin According To The Redundancy Analysis.

Table 6. The Rda Results For The Percentage Of The Overall River Water Quality Variance Explained By An Individual Variable And All The Explanatory Variables

Explanatory Variables	Variance Explained%	P Value
Land Use		
Frt	29.6	0.002
Wt	8.8	0.16
Ub	50.8	0.002
Crp	16.2	0.016
Topography		
Elv	29.8	0.004
Slp_{mean}	36.3	0.002
Ar	15.7	0.032
Socio-Economic Factor		
Hp_d	44.6	0.002
Gdp_{pc}	26.2	0.032
Aho_{pc}	26.9	0.002
Land Use + Topography + Social * Economic Factors	86.1	0.002

Table 7. The Rda Results Showing The Eigenvalues And Percentage Of The Overall River Water Quality Parameter Variance Explained By The First Two Rda Axes And All The Rda Axes

	Axis 1	Axis 2	All Axes
Eigenvalues	0.524	0.165	0.861
Percentage Of Water Quality Variance Explained	52.4	16.5	86.1
Dominant Environmental Factors	Ub(0.967)	Aho_{pc}(-0.699)	
	Hp_d (0.897)		
	Slp_{mean}(-0.780)		
	Frt(-0.725)		

DISCUSSION

Seasonal Effects

Seasonal variations in the flow caused by the subtropical monsoon climate can partially explain the river water quality temporal variations [7], [38]. The relatively high TN concentrations in the rainy season were attributed to a flushing effect [36]. BOD_5, which corresponds to point source pollution, were lower in the rainy season due to the dilution by precipitation [2]. Relatively high T in the rainy season resulted from the specific climatic conditions during this period (T is higher in the rainy season than in the dry season; see **Table 2**). The lower values of DO and higher values of pH in the rainy season may related to some environmental factors variations, such as flow rate, water temperature, aquatic plant growth and so on [50], [39]–[41]. The reasons are complex and the further research is needed in the future. The turbidity was higher in the dry season, which may be primarily because the reduced flow during this period caused the river water to be easily influenced by tides. Moreover, Odokuma and Okpokwasili [40] reported that the other parameters such as phosphate showed significantly higher values in the rainy season than in the dry season in the New Calabar River, Nigeria. These different results compared with our study were mainly due to the regional climate difference. However, the absence of a significant difference in the other physico-chemical parameters between the dry and wet seasons indicated mixed and irregular influences, e.g., point sources and diffusion may have played an important role [1], [42].

Land cover Effects

In the study area, land-use types could play important roles in affecting major river water quality parameter distributions, as reflected by their considerable variability (**Figure 2**). This result is corroborated by the strong positive correlations between urban land and BOD_5, COD_{Mn}, TN, DN, NH_4^+-N and TP (**Table 5**) and the strong negative correlations with DO and pH. Moreover, most of the nutrient variables exhibited lower concentrations in the forest-dominated region, whereas the cropland-dominated region had high nutrient concentrations. The region downstream of the urban area exhibited higher concentrations of BOD_5, COD, NH_4^+-N, TP and DP and lower pH and DO (**Figure 2**); this finding agrees with many related studies [21], [43], [44]. Studies have also shown that certain river water quality parameters (e.g., TN and TP) are primarily determined by agricultural land use at the watershed scale in many parts of the world [1], [20]. However, this study found that agricultural land use in watersheds plays a minor role in explaining the spatial variations in river water quality (explaining 16.2% of the total overall water quality variance) in the Cao-E River basin. The different result is primarily due to the enhanced effects of urban runoff and incompletely treated industrial waste and domestic pollution on river water quality compared to agricultural runoff. In addition, vegetated buffers adjacent to the croplands can substantially mitigate agricultural runoff of nutrients and other contaminants via deposition, absorption, and denitrification [20].

Urban areas are primarily located along river networks in the Cao-E River basin. Therefore, urban effects on the river water quality were expected. Urban land use comprised a much smaller percentage of the Cao-E River basin than the cropland in this study. However, as observed from the regression model (BOD_5, COD_{Mn}, TN, DN, NH_4^+-N, NO_3^--N, TP, DO and pH) and the RDA analysis (explaining 50.8% of the total overall water quality variance) results, urban areas play a pivotal role in influencing water quality (**Tables 5 and 6**). This observation is most likely due to two factors: (1) the deficient capacity of urban sewage treatment plants leads to domestic pollution and industrial wastewater inputs into the Cao-E River system, and (2) the high percentage of impervious surfaces and over fertilization of grassy areas can increase discharge rates, sedimentation and pollutant runoff to streams [44].

Topography Effects

Topographical factors played important roles in explaining spatial variations in river water quality within the Cao-E River basin (SLP_{mean} was 36.3% and ELV was 29.8%). Topography largely regulated the river water quality parameters. Most related studies have shown that higher SLP_{mean} and/or ELV lead to higher erosion rates, which subsequently increase the rate at which particulate matter enters a water body [13], [21], [45]. However, our results indicated an inverse relationship. Here, SLP_{mean} and ELV were both negatively correlated with TN, DN, NO_3^--N, TP and DP; SLP_{mean} exhibited a highly significant negative correlation with NO_3^--N values (**Figure 3 and Table 5**). The slope effects on water chemistry varied. Watershed physical characteristics, such as soil properties (soil texture and soil drainage), morphological variables (drainage density and elongation) and particularly surficial debris, largely affected water chemistry in river waters [37], [46]. In addition, these negative correlations existed primarily because ELV and SLP_{mean} were strongly correlated with land cover (**Table 8**). High elevations and/or steep catchments are primarily occupied by forest land cover; however, cropland is primarily located in regions in which the topography is relatively flat and has a low ELV. Therefore, the forested mountain area, which has higher SLP_{mean} and/or ELV, may export fewer nutrients than flat land (e.g., cropland). Similar results were published regarding water quality spatial variations in the Han River basin in South Korea [7] and non-point source effects on stream nutrient concentrations in the Seattle region of the USA [47].

Table 8. The Pearson Correlation Matrix For The Watershed Characteristics Of The Cao-E River Basin In Eastern China.

	Frt	Wt	Ub	Crp	Ely		Ar	^{11}Po	$Gdpp_c$	Aho"
Frt	1.00									
Wt	0.09	1.00								
Ub	-0.79**	0.31	1.00							
Crp	-0.96**	-0.31	0.59**	1.00						
Elv	0.57**	-0.48**	-0.71**	0.41	1.00					
Slp_{mean}	0.83**	-0.33	-0.82**	0.70**	0.74**	1.00				
Ar	-0.44	0.17	0.42	0.37	-0.44	-0.56*	1.00			
Hpd	-0.79**	0.37	0.95**	0.60	-0.79**	-0.91**	0.55*	1.00		
Gdp_{pc}	-0.38	0.49*	0.69**	0.18	-0.65**	-0.67**	0.30	0.76**	1.00	
Aho_{pc}	-0.47*	0.20	0.55*	0.36	-0.69**	-0.72**	0.47**	0.64**	0.72**	1.00

The effects of socio-Economic Factors

The socio-economic factors commonly have different extent effects on river water quality, especially in developing countries such as China. Our study shows that socioeconomic factors have significant effects on river water quality during the study period (**Table 5**) and that HP_d, AHO_{pc}, and GDP_{pc} explain 44.6%, 26.9% and 26.2% of the total overall river water quality variance, respectively (**Table 6**). Furthermore, HP_d is a fundamental parameter for predicting DP and AHO_{pc} for TP and turbidity (**Table 5**). This result can be attributed to high population density, the massive animal production and the lack of strict management regulations for animal waste in the study area. The population density exceeds 380 people km^{-2}, the annual animal output value exceeds 2.0 billion Yuan and the annual hog production exceeds 1.0 million heads for the entire basin [48]. Most human and animal waste with incomplete treatment is discharged into the surrounding water bodies. Moreover, EC was positively correlated with GDPpc, which was most likely related to point source pollution [37].

In addition, some researches [4]–[6] point out that climate condition is also an important factor of influencing river water quality. Thus, we will further study it in our future work.

CONCLUSIONS

The primary results of this study can be summarized as follows:

- The in-stream water quality in the Cao-E River basin streams suggested that TN, pH and T were generally higher in the rainy season, whereas BOD_5, DO and turbidity were higher in the dry season.
- Spatial variations in river water quality are typically associated with several anthropogenic and natural factors. Urban land cover was determined to be the most important explanatory variable for BOD_5, COD_{Mn}, TN, DN, NH_4^+-N, NO_3^--N, TP, DO and pH. Moreover, Animal husbandry output per capita was an important predictor for TP and turbidity, and Gross domestic product per capita largely determined the spatial variations in EC. The remaining unexplained variance resulted from other factors, such as topography.

- Pollution control for animal waste discharge in rural settlements was important in the study area. Moreover, agricultural runoff, industrial pollution and domestic pollution in urban and industrial areas were also important factors within the Cao-E River basin.
- The percentage of total overall river water quality variance explained by individual variables and/or all environmental variables in the study area (as determined using RDA) can assist in quantitatively identifying primary pollution control factors at the watershed scale.

This study improved our understanding of the anthropogenic activities and natural factors that affect river water quality and can assist in the design of efficient strategies for controlling river water pollution at the watershed scale. Moderate-resolution remote sensing data were adopted in this study; future investigations will require high-resolution DEM maps and additional land use classes to better evaluate the effects of specific environmental variables on overall river water quality.

ACKNOWLEDGMENTS

We would like to acknowledge the members of the Zhejiang University soil physical laboratory for assistance in the field and laboratory.

REFERENCES

1. Li S, Gu S, Liu W, Han H, Zhang Q (2008) Water Quality In Relation To Land Use And Land Cover In The Upper Han River Basin, China. Catena 75: 216–222. doi: 10.1016/j.catena.2008.06.005
2. Zhang Y, Guo F, Meng W, Wang X (2009) Water Quality Assessment And Source Identification Of Daliao River Basin Using Multivariate Statistical Methods. Environ Monit Assess 152: 105–121. doi: 10.1007/s10661-008-0300-z
3. Zhou N, Westrich B, Jiang S, Wang Y (2011) A Coupling Simulation Based On A Hydrodynamics And Water Quality Model Of The Pearl River Delta, China. J Hydrol 396: 267–276. doi: 10.1016/j.jhydrol.2010.11.019

4. Astaraie-Imani M, Kapelan Z, Fu G, Butler D (2012) Assessing The Combined Effects Of Urbanisation And Climate Change On The River Water Quality In An Integrated Urban Wastewater System In The Uk. J Environ Manage 112: 1–9. doi: 10.1016/j.jenvman.2012.06.039
5. Costa-Cabral M, Coats R, Reuter J, Riverson J, Sahoo G, et al. (2013) Climate Variability And Change In Mountain Environments: Some Implications For Water Resources And Water Quality In The Sierra Nevada (Usa). Clim Change 116: 1–14. doi: 10.1007/s10584-012-0630-2
6. Aubert Ah, Gascuel-Odoux C, Merot P (2013) Annual Hysteresis Of Water Quality: A Method To Analyse The Effect Of Intra-And Inter-Annual Climatic Conditions. J Hydrol 478: 29–39. doi: 10.1016/j.jhydrol.2012.11.027
7. Chang H (2008) Spatial Analysis Of Water Quality Trends In The Han River Basin, South Korea. Water Res 42: 3285–3304. doi: 10.1016/j.watres.2008.04.006
8. Hoorman J, Hone T, Sudman Jr T, Dirksen T, Iles J, et al. (2008) Agricultural Impacts On Lake And Stream Water Quality In Grand Lake St. Marys, Western Ohio. Water Air Soil Pollut Int 193: 309–322. doi: 10.1007/s11270-008-9692-1
9. Ahearn Ds, Sheibley Rw, Dahlgren Ra, Anderson M, Johnson J, et al. (2005) Land Use And Land Cover Influence On Water Quality In The Last Free-Flowing River Draining The Western Sierra Nevada, California. J Hydrol 313: 234–247. doi: 10.1016/j.jhydrol.2005.02.038
10. Huang F, Wang X, Lou L, Zhou Z, Wu J (2010) Spatial Variation And Source Apportionment Of Water Pollution In Qiantang River (China) Using Statistical Techniques. Water Res 44: 1562–1572. doi: 10.1016/j.watres.2009.11.003
11. Chang N (2005) Sustainable Water Resources Management Under Uncertainty. Stoch Environ Res Risk Assess 19: 97–98. doi: 10.1007/s00477-004-0217-1
12. Kolovos A, Christakos G, Serre Ml, Miller Ct (2002) Computational Bayesian Maximum Entropy Solution Of A Stochastic Advection - Reaction Equation In The Light Of Site - Specific Information. Water Resour Res 38: 51–54. doi: 10.1029/2001wr000743

13. Sliva L, Dudley Williams D (2001) Buffer Zone Versus Whole Catchment Approaches To Studying Land Use Impact On River Water Quality. Water Res 35: 3462–3472. doi: 10.1016/s0043-1354(01)00062-8
14. Heathcote Aj, Filstrup Ct, Downing Ja (2013) Watershed Sediment Losses To Lakes Accelerating Despite Agricultural Soil Conservation Efforts. Plos One 8: E53554. doi: 10.1371/journal.pone.0053554
15. Keatley Be, Bennett Em, Macdonald Gk, Taranu Ze, Gregory-Eaves I (2011) Land-Use Legacies Are Important Determinants Of Lake Eutrophication In The Anthropocene. Plos One 6: E15913. doi: 10.1371/journal.pone.0015913
16. Wei R, Ge F, Huang S, Chen M, Wang R (2011) Occurrence Of Veterinary Antibiotics In Animal Wastewater And Surface Water Around Farms In Jiangsu Province, China. Chemosphere 82: 1408–1414. doi: 10.1016/j.chemosphere.2010.11.067
17. Xianghao Z (2008) Study Of Protection And Construction Of Mountain Ecological Security Barrier In China. J Mount Sci (In Chinese) 1: 4–5.
18. Jiang X, Sommer Sg, Christensen Kv (2011) A Review Of The Biogas Industry In China. Energy Policy 39: 6073–6081. doi: 10.1016/j.enpol.2011.07.007
19. Tong L, Li P, Wang Y, Zhu K (2009) Analysis Of Veterinary Antibiotic Residues In Swine Wastewater And Environmental Water Samples Using Optimized Spe-Lc/Ms/Ms. Chemosphere 74: 1090–1097. doi: 10.1016/j.chemosphere.2008.10.051.
20. Liu W, Li S, Bu H, Zhang Q, Liu G (2012) Eutrophication In The Yunnan Plateau Lakes: The Influence Of Lake Morphology, Watershed Land Use, And Socioeconomic Factors. Environ Sci Pollut Res Int 19: 858–870. doi: 10.1007/s11356-011-0616-z
21. Ye L, Cai Q, Liu R, Cao M (2009) The Influence Of Topography And Land Use On Water Quality Of Xiangxi River In Three Gorges Reservoir Region. Environ Geol 58: 937–942. doi: 10.1007/s00254-008-1573-9
22. Tian C, Lu X, Pei H, Hu W, Xie J (2013) Seasonal Dynamics Of Phytoplankton And Its Relationship With The Environmental Factors In Dongping Lake, China. Environ Sci Pollut Res Int 185: 2627–2645. doi: 10.1007/s10661-012-2736-4

23. Yang X, Wu X, Hao H, He Z (2008) Mechanisms And Assessment Of Water Eutrophication. J Zhejiang Univ Sci B 9: 197–209. doi: 10.1631/jzus.b0710626

24. .Chen D, Lu J, Shen Y, Gong D, Deng O (2011) Spatio-Temporal Variations Of Nitrogen In An Agricultural Watershed In Eastern China: Catchment Export, Stream Attenuation And Discharge. Environ Pollut 159: 2989–2995. doi: 10.1016/j.envpol.2011.04.023

25. Jin S, Lu J, Chen D, Shen Y, Shi Y (2009) Relationship Between Catchment Characteristics And Nitrogen Forms In Cao-E River Basin, Eastern China. J Environ Sci (China) 21: 429–433. doi: 10.1016/s1001-0742(08)62287-1

26. Shen Y, L U J, Chen D, Shi Y (2011) Response Of Stream Pollution Characteristics To Catchment Land Cover In Cao-E River Basin, China. Pedosphere 21: 115–123. doi: 10.1016/s1002-0160(10)60086-0

27. Sepbc (2002) Water And Wastewater Analysis Method. Prepared And Published By: State Environment Protection Bureau Of China (Sepbc). Beijing: China Environmental Science Press. 784 P.

28. Dodds Wk, Oakes Rm (2004) A Technique For Establishing Reference Nutrient Concentrations Across Watersheds Affected By Humans. Limnol Oceanogr Methods 2: 333–341. doi: 10.4319/lom.2004.2.333

29. Chen J, Lu J (2013) Establishment Of Reference Conditions For Nutrients In An Intensive Agricultural Watershed, Eastern China. Environ Sci Pollut Res Int 4: 2496–2505. doi: 10.1007/s11356-013-2166-z

30. Olsen Rl, Chappell Rw, Loftis Jc (2012) Water Quality Sample Collection, Data Treatment And Results Presentation For Principal Components Analysis–Literature Review And Illinois River Watershed Case Study. Water Res 46: 3110–3122. doi: 10.1016/j.watres.2012.03.028

31. Myers Rh (1990) Classical And Modern Regression With Applications: Duxbury Press Belmont, Ca. 350p.

32. 32.Mial Miserendino, Casaux R, Archangelsky M, Di Prinzio Cy, Brand C, et al. (2011) Assessing Land-Use Effects On Water Quality, In-Stream Habitat, Riparian Ecosystems And Biodiversity

In Patagonian Northwest Streams. Sci Total Environ 409: 612–624. doi: 10.1016/j.scitotenv.2010.10.034

33. Braak Ct, Šmilauer P (2002) Canoco Reference Manual And Canodraw For Windows User›s Guide: Software For Canonical Community Ordination (Version 4.5). Section On Permutation Methods. Microcomputer Power, Ithaca, New York. 300p.
34. Lepš J, Šmilauer P (2003) Multivariate Analysis Of Ecological Data Using Canoco. New York: Cambridge University Press. 283 P.
35. Johnson Ra, Wichern Dw (2002) Applied Multivariate Statistical Analysis: Prentice Hall Upper Saddle River, Nj. 755p.
36. Chen D, Lu J, Yuan S, Jin S, Shen Y (2006) Spatial And Temporal Variations Of Water Quality In Cao-E River Of Eastern China. J Environ Sci (China) 18: 680–688.
37. Li S, Xia X, Tan X, Zhang Q (2013) Effects Of Catchment And Riparian Landscape Setting On Water Chemistry And Seasonal Evolution Of Water Quality In The Upper Han River Basin, China. Plos One 8: E53163. doi: 10.1371/journal.pone.0053163
38. Shah Vg, Dunstan Rh, Geary Pm, Coombes P, Roberts Tk, et al. (2007) Comparisons Of Water Quality Parameters From Diverse Catchments During Dry Periods And Following Rain Events. Water Res 41: 3655–3666. doi: 10.1016/j.watres.2007.02.052
39. Karaer F, Küçükballi A (2006) Monitoring Of Water Quality And Assessment Of Organic Pollution Load In The Nilüfer Stream, Turkey. Environ Monit Assess 114: 391–417. doi: 10.1007/s10661-006-5029-y
40. Odokuma Lo, Okpokwasili Gc (1997) Seasonal Influences Of The Organic Pollution Monitoring Of The New Calabar River, Nigeria. Environ Monit Assess 45: 43–56. doi: 10.1007/bf02394664
41. Kannel Pr, Lee S, Lee Y (2008) Assessment Of Spatial–Temporal Patterns Of Surface And Ground Water Qualities And Factors Influencing Management Strategy Of Groundwater System In An Urban River Corridor Of Nepal. J Environ Manage 86: 595–604. doi: 10.1016/j.jenvman.2006.12.021
42. Chen Js, Gao Xm, He Dw, Xia Xh (2000) Nitrogen Contamination In The Yangtze River System, China. J Hazard Mater 73: 107–113.

43. Pratt B, Chang H (2012) Effects Of Land Cover, Topography, And Built Structure On Seasonal Water Quality At Multiple Spatial Scales. J Hazard Mater 209: 48–58. doi: 10.1016/j.jhazmat.2011.12.068
44. Mehaffey Mh, Nash Ms, Wade Tg, Ebert Dw, Jones Kb, et al. (2005) Linking Land Cover And Water Quality In New York City›s Water Supply Watersheds. Environ Monit Assess 107: 29–44. doi: 10.1007/s10661-005-2018-5
45. Xiao B, Wang Q, Wu J, Huang C, Yu D (2010) Protective Function Of Narrow Grass Hedges On Soil And Water Loss On Sloping Croplands In Northern China. Agric Ecosyst Environ 139: 653–664. doi: 10.1016/j.agee.2010.10.011
46. Lindell L, Aa Str O M M, O Berg T (2010) Land-Use Change Versus Natural Controls On Stream Water Chemistry In The Subandean Amazon, Peru. Appl Geochem 25: 485–495. doi: 10.1016/j.apgeochem.2009.12.013
47. Brett Mt, Arhonditsis Gb, Mueller Se, Hartley Dm, Frodge Jd, et al. (2005) Non-Point-Source Impacts On Stream Nutrient Concentrations Along A Forest To Urban Gradient. Environ Manage 35: 330–342. doi: 10.1007/s00267-003-0311-z
48. Shaoxing City Statistical Bureau (2005) Shaoxing Statistical Yearbook. Shaoxing: Shaoxing City Press. 300p.
49. Belsey Da, Kuh E, Welsch Re (1980) Regression Diagnostics: Identifying Influential Data And Sources Of Collinearity: John Wiley. 320p.
50. Zhao J, Fu G, Lei K, Li Y (2011) Multivariate Analysis Of Surface Water Quality In The Three Gorges Area Of China And Implications For Water Management. J Environ Sci (China) 23: 1460–1471. doi: 10.1016/s1001-0742(10)60599-2

Chapter 3

Stakeholder Involvement in Sustainable Watershed Management

Reyhan Erdogan

[1]Akdeniz University, Agricultural Faculty, Department of Landscape Architecture, Antalya, Turkey

INTRODUCTION

Water is a Main Source of Life

Water is a key element for human life and philosophy. Many of the thinkers had considered on this issue and they have produced several theories. For example, Thales' most famous belief was his cosmological thesis, which held that the world started from water. This thesis was the first try to explain how was the beginning of the creation and world, in the terms of the logical and systematical way. On the other side, Aristotle considered this belief roughly equivalent to the later ideas of Anaximenes, who held that everything in the world was composed of air. The best explanation of Thales' view is the following passage from Aristotle's Metaphysics (Anonymous, 2012a). The passage contains words from the theory of matter and form that were adopted by science with quite different meanings. "That from which is everything that exists and from which it first becomes and into which it is rendered at last, its substance remaining under it, but transforming in qualities, that they say is the element and principle of things that are." Again: "For it is necessary that there be some nature, either one or more than one, from which become the other things of the object being saved... Thales the founder of this type of philosophy says that it is water". A deeper dip into the waters of the theory of matter and form is properly reserved to other articles (Anonymous, 2012a). Thales taught as follows: "Water constituted the principle of all things." Heraclitus Homericus states that Thales drew his conclusion from seeing moist substance turn into air, slime and earth. It seems likely that Thales viewed the Earth as solidifying from the water on which it floated and which surrounded Ocean (Anonymous, 2012b). On the other hand, Heraclitus said, "everything in life is changing and will change too". He is famous for his insistence on ever-present change in the universe, as stated in the famous saying, "No man ever steps in the same river twice". This quote is the famous speech about waters and river in the philosophy.

On the other hand, problems about the water should be necessarily solved and it is very important for human life. The main problem about water is permanent since the human history. In the right term humans are the main factor of the problems about water and rivers, they have been effected by these problems in the first hand, but in the same time humans are responsible about to solve these problems The most challenging activity about water is the shape and nature of the water. Water can break walls, buildings and cities. Besides this destructive nature of the water, humans need water for sustaining their lives too. They need it, to irrigate their fields for producing crops, to drink when they are thirsty. For solving this, problems humans need to improve their calculations and they try to find a suitable solution for these situation. A first innovative and qualitative solution in math has been established in this period, in the field of the geometry (Deauna et al., 2000). Because of them the earliest branches of Mathematics. Beginnings of the geometry can be traced before 1700 B.C. or in early ancient Egypt. Due to necessity, every time the Nile River inundated and deposited fertile soil along the bank, the early Egyptian had to solve the problem of size and boundaries of land along the Nile River. Changes happened in the contour of the land had caused confusion among landowners. Therefore, a system of making boundaries, measuring lengths and areas had to be discovered (Deauna et al., 2000). From this circumstance, the name "Geometry" has evolved. The word "Geometry" originated from the word "Geo" means "earth" in Greek and "metros" means "to measure" (Deauna et al., 2000).

Many thinkers of the western World could not be stand indifferent to these rules of the nature. They were very proud to these knowledge and they have adopted these knowledge to their societies. These situation is very ironic because today many of the eastern countries are following western World. However, nowadays the source of the knowledge is not matter; every country and every nation in the World want to solve the water problems. Water is the most important source in the World, so water problem is now a global problem and it is a key issue for every individual in the World.

Watershed

A watershed is an area of land where all of the water that is under it or drains off it goes into the same place. John Wesley Powell, geographer and a scientist, put it best when he said that a watershed is: "that an area of land, a bounded hydrologic system, within which all living things are inextricably linked by their common water course and where, as humans settled, simple logic demanded that they become part of a community." (USEPA, 2012). The term "watershed" is commonly used to describe an area of the earth's surface from which water flows downhill to a single outflow point (Anonymous, 2009). Watersheds come in all shapes and size. The area encompassed may either be small, such as that which an ephemeral stream drains only during precipitation events, or be large. Watersheds cross county, state, and national boundaries (Anonymous, 2009). Because of them managing of the watersheds is very sensitive political issue between countries.

Health of the watershed has directly effect on the quality of the water, which drain in the watershed area. In this situation, quality of water is important (Giri et.al,. 2012). Furthermore, the size and volume of the water are other important indicators of the water management. Maintaining proper water quality conditions is important to protect human, animal, and plant health, and is an on going concern in water resources (Giri et.al,. 2012). Meanwhile, anthropogenic activities such as direct industrial discharges and agricultural practices significantly interfere with natural processes, which ultimately degrade water quality (Giri et.al,. 2012). Natural systems are mainly working on the logic of life cycle. A change in a parameter can be effect all other parameters.

Water quality depends on several physical combinations (Said et al., 2006). Sustainability of these physical components is important. Basically, physical sustainability is the combination of water quality, water quantity and ecological factors (Said et al., 2006). Evaluating the water quality situation within a watershed requires information on physical, chemical, and biological water quality indicators. Unfortunately, for many constituents historical measurements do not exist for most watersheds. In these cases, estimation of these constitutes is necessary from the available data on different, related constituents (Said et al., 2006). Trends of these constituents are very important for complete evaluation and mitigation of water quality problems and in the long-term decision-making processes. Using advanced techniques that construct relationships between water quality constituents makes evaluation and mitigation of water quality problems more reliable (Said et al., 2006).

There are some international tools to protect and improve water quality. One of these tools is European water Framework. This framework is a proper way to combine countries for more sustainable management practices. There are many good examples of the plans that aim to achieve a better understanding of water bodies. To comply with the European Water Framework Directive (EWFD, 2000/60/EC), which requires a "Good Ecological Status" for all waterbodies in 2015, French policy adopted a Phyto-Pharmaceutic Products' or pesticide reduction plan (ECOPHYTO 2018 plan). This program aims at inciting farmers to halve the annual amount of pesticides used. Nevertheless, the question of pesticide transfer is not solved. Presently the "zero pesticide" solution is not achievable for technical and economical reasons. The sociological part of the study highlights the importance of a co-construction process between all involved stakeholders, even if it is time consuming. Theoretical solution should be adapted not only to hydrological aspects but also to the socio-economical context (Tournebize et al., 2011).

Altering the landscape can influence the hydrologic integrity of a watershed. Changes in water quality and quantity that can occur from storm-water runoff include but are not limited to:

- Decreased and/or lack of infiltration to groundwater recharge areas and potential aquifer depletion,
- Decreased seasonal stream base-flows,
- Increased storm water volume and velocity contributing to stream-bank erosion, stream channelization, and more frequent and severe flood events,
- Increased surface water temperatures from heated storm-water runoff and lack of shaded riparian buffers, and
- Decreased filtration of pollutants from narrowing or damaging riparian buffers (Tournebize et al., 2011).

Mitigation of water quality problems requires a number of actions by stakeholders and regulatory agencies. In some cases, especially in locally impaired waters, stakeholders can identify and mitigate water quality problems directly. In the case of significant local or regional pollution, watershed management plans and the total maximum daily loads, issued by USEPA of the Clean Water Act, enhance the water quality situation in impaired watersheds (USEPA, 1991). The total maximum daily load establishes a set of required reductions in target parameters consistent with an implementation policy to protect in stream water quality (Benham et al., 2002). If the water quality continues to worsen or goals can not be achieved under easing condition, it may be necessary to apply point source reductions or best management practices to improve the water quality as a whole. In summary, the water quality component in watershed management integrates scientific knowledge and stakeholder supplied information regarding the watershed-scale social, economic, and environmental processes affecting water quality.

Watershed Management

Watershed management is the integrated use of land, vegetation and water in a geographically discrete drainage area for the benefit of its residents, with the objective of protecting or conserving the hydrologic services that the watershed provides and of reducing or avoiding negative downstream or groundwater impacts (Darghouth et al., 2008).

Watershed management is dedicated to solving watershed problems on a sustainable basis. These problems can be classified into five categories: lack of water (quantity), deterioration in water quality, ecological affects, poor public participation, and low output economic value for the investment in water-related activities. The first three categories constitute physical or "natural" sustainability while the last two categories define social and economic sustainability. For successful implementation of solutions to the physical and economic issues, a broad, representative array of stakeholders should be involved (Grigg, 1998, Said et al., 2006).

Watersheds are a highly desirable unit for planning because they are physical features ubiquitous across the landscape serving as the geographic foundation for political states. As planning units, watersheds transcend political boundaries. However, prior to the 1970's, most watershed management focused on solving localized problems without taking into account the interrelationship between those problems and the biophysical, economic and social elements of the larger watershed system (Heathcote, 1998). Furthermore, during most of the mid- to late- 20th century, watershed management was, politically, a top-down planning process with national concern pre-empting local (National Research Council, 1999). In the mid-1990s, EPA's Risk Assessment Forum and Office of Water cosponsored the development of five-demonstration watershed ERAs to test application of the ERA Framework (USEPA, 1992) to Office of Water programs. The five watershed assessments were performed in the Clinch and Powell Valley, VA, Middle Snake River, ID, Waquoit Bay, MA, Big Darby Creek, OH, and the Middle [segment of the] Platte River, NE. All five selected sites had valued ecological resources, multiple stressors, an existing data set, and willing assessment participants. Using these assessments, many key aspects of the watershed ERA process, including challenges encountered and lessons learned, documented.

Chess and Gibson (2001) stated that the attributes intrinsic to a watershed management are: scientific feasibility, social feasibility, and motivational feasibility. They considered that these feasibilities are not static attributes although they can be improved through additional efforts to deal with attributes that limit feasibility. Watershed management plans may not be implemented due to lack of support even if the planning process and proposed actions are faultless. However, watershed restoration and protection management are often a complex web of technical, legal, and economic issues involving numerous stakeholders. The ability to address these issues depends on the attributes intrinsic to the watershed and whether they provide sufficient, basis for this innovative approach to watershed management (Chess and Gibson 2001). These attributes are defined by Chess and Gibson (2001) as follows:

- Scientific feasibility is tied to the nature of environmental problems and the scientific capacity to solve them.
- Social feasibility encompasses public communications and engagement.
- Motivational feasibility includes issues related to values or economic consideration (Chess and Gibson 2001).

Scientific feasibility concerns the nature of the environmental problems, the tools to track them, and the methods to improve them. Examples of scientific feasibility related issues are the engineering solutions to improve water quality or conserve water quantity, technological limitations, such as the inability to detect the source of pollution, the absence of obvious causal relationships, and the development of best management practices (Said et al., 2006). In some areas, scientific feasibility may be so limited that management of a given watershed becomes a lower priority than other watersheds in which watershed management is more likely to succeed. Research and strong leadership can increase scientific feasibility to overcome technological barriers (Said et al., 2006). In summary, environmental problems must be measurable and tractable over time in order to predict performance, achieve public awareness, indicate progress towards goals, locate hot spots, take actions when appropriate, and sustain watershed management (Said et al., 2006).

Social feasibility means getting stakeholders involved and utilizing their input in a meaningful process. The National Research Council noted the need for watershed management to "get the participation right." There are many ways to improve stakeholder partnerships (Said et al., 2006). These include recruiting participants, encouraging leadership, education, civic engagement, and exploring various structural arrangements. Watershed management can be implemented through voluntary participation or facilitated by existing statutory infrastructure. Ideally, all watershed management activities would be driven by voluntary participation because of enlightened self-interest. However, statutory authority is often important because of the authority it grants, the planning it may mandate, and the funding it may provide (Said et al., 2006). Collaborative approaches may not be appropriate for watersheds in which there is limited evidence of social feasibility. If there is lack of social feasibility within a watershed, Federal and state agencies need to recognize that significant effort may be required to build social capacity to compensate for this intrinsic limitation. Agencies can build social feasibility by promoting civic efforts and by following best practices for development of collaborations (Said et al., 2006).

Occurrences of flooding, drinking water contamination, or other threats may increase social feasibility. Motivational feasibility is needed when there is a lack of sufficient income or demographic composition and when tax and donation of funds and time are needed. Motives may fall into three distinct categories: rational (cost-benefit assessment), norm-based (efforts to conform to norms such as pollution remediation), and affective (emotional attachments that are difficult to quantify such as protection of endangered species and non-use values) but are powerful in watershed management. Motivational feasibility is closely linked to social feasibility. For watershed management activities to occur there must be motivation to take action. Motivation can be induced by infusing funds (incentives for watershed projects), increasing values, and improving cost benefit ratios (Said et.al, 2006).

There is some important plan at maintaining progress towards a sustainable watershed. The first one is organizational capacity for a long term. The second one is multi-disciplined, inclusive oversight group. The others are revitalization of existing planning efforts and programs as planning tools for a sustainable future; broadening the scale of future sustainability efforts in the region to include the entire Broad-head watershed; further review and analysis of the regulatory framework affecting sustainable watershed planning efforts, and continuation of an education and outreach effort through social marketing efforts.

Identifying and Including Stakeholders in Watershed Management

Among the various ways of identifying stakeholders, as well as in the agency, behavioral, ecological, institutional, resource dependence, and transaction cost theories of the firm, there has been found no single attribute within a given theory that can guide people to a reliably on these issues. However, it was found that one can extract from these literatures the idea that just a few attributes can be used to identify different classes of stakeholders in an environment. Analysis with Freeman's definition of stakeholder is a good point. He stated that-"any group or individual who can affect or is affected by the achievement of the organization's objectives" (Freeman, 1984) and develop a theory of stakeholder identification drawn from these various theoretical literatures. From this point, there can be a broad definition can be explained, so that no stakeholders, potential or actual, are excluded from analysis arbitrarily or a priori (Wamalwa, 2009). Stakeholders are important in the planning of the natural environment where the human activities and natural processes are multi-layered combined. And then, a propose that classes of stakeholders can be identified by their possession or attributed possession of one, two, or all three of the following attributes (Wamalwa, 2009):

- the stakeholder's power to influence life on the watershed,
- the legitimacy of the stakeholder's relationship with the watershed region, and
- the urgency of the stakeholder's claim on the watershed area (Wamalwa, 2009).

This theory produces a comprehensive typology of stakeholders based on the normative assumption that these variables define the field of stakeholders: those entities to whom managers should pay attention (Wamalwa, 2009). This attention is a very important key factor that the sustainability of the watershed management.

According to Mitchell et al., (1997) stakeholders can be classified as "Latent" referring to those who are of low importance to the organization, because they posses only one of the three characteristics (urgency, power and legitimacy) or "Expectant", having a higher importance, than the latent by virtue of being in possession of two of the three qualities or "definitive" having the highest degree of importance to the organization, because they posses all the three features. The latent group of stakeholders can be further classified into three (Mitchell et. al. 1997). Dormant; involving those who have power but of no or little effect to the goal of integration, because they do not have legitimacy and urgency. Discretionary; including those who have legitimacy but have no urgency and power. Demanding; including those who may have the urgency but of low rating because they do not have legitimacy and power (Mitchell et. al. 1997). The expectant group can also be further subdivided into dominant, wielding both high influence and stake or, dependent with high stake and urgency but having low influence, or dangerous with urgency and power (Wamalwa, 2009).

The planning phase for a watershed assessment is especially complex because a watershed typically overlaps multiple jurisdictions that are managed by organizations with divergent goals and responsibilities and inhabited by numerous stakeholders with varied interests. Depending on the perspectives of the stakeholders involved, a different set of goals and objectives may emerge. Stakeholder participation may become complex if there are numerous stakeholders or some of them are interested in only a limited range of issues (Glicken, 2000).

The inadvertent exclusion of a stakeholder group may influence a group's decision to accept or reject the outcome of the process. Excluded stakeholders could even take legal actions, such as filing a citizen lawsuit under the Clean Water Act to challenge the process outcomes. Therefore, stakeholder involvement needs to be balanced against the limited resources available for watershed assessment and management (USEPA, 2001).

Participants in the watershed management process should include all regulatory or resource agencies (state, federal, local) with responsibilities for protecting and managing the water body and any parties whose authority will be needed to implement a management plan (e.g., local and county officials with zoning oversight). Nongovernmental organizations (such as watershed associations or councils, river watch citizen groups, volunteer monitoring group, educational and research institutions, industries, and agricultural associations) all have a stake in watershed management. Besides organized groups, other stakeholders include landowners, those who use the watershed, and those whose participation is essential to successful management. In some instances, stakeholders may be hundreds of miles away from the assessment (e.g., bird watchers concerned about migratory waterfowl). The nongovernmental and unaffiliated stakeholders may have objectives that are very different from those of the regulatory agencies (e.g., minimizing restrictions on land use, resource development, or waste disposal). Although such social, legal, and economic objectives may be in conflict with some environmental objectives, they are still relevant concerns that need to be considered (Stahl et al., 1999). To help assess these tradeoffs, it may be helpful to involve environmental economists.

Participants on a watershed assessment team contribute different resources that include socioeconomic information, historical data, scientific expertise, and assets to conduct the assessment. The Big Darby Creek assessment project is a good example of team studies in the watershed management. Team members of the Big Darby Creek assessment project included biologists, a city planner, and environmental scientists (Cormier et al., 2000). Team members performed literature reviews and frequently consulted or interviewed experts in other disciplines. The team interacted regularly with the Big Darby Partners, a group of state agencies, representatives from The Nature Conservancy, and farmers concerned about the future of the working landscape. Because of them partnership between stakeholders and planners are important factor. Integration between these two groups effect the plan and application.

WATERSHED MANAGEMENT APPROACHES

Collaborative Watershed Management

Collaborative watershed management has emerged in the last two decades as a promising approach to address non-point source pollution in waters. With such a wide variety of land-use patterns across watersheds, it is important that collaborative approaches to water resource management are tailored to local land-use planning efforts (Wang, 2001; Scott et. al., 2010). Urban and rural landscapes can have very different biological systems, leading watershed partnerships located in different areas to address different environmental issues. Moreover, collaborative management efforts in each setting can be impacted by different sets of variables, from the level of human capital (e.g., income, education) and social capital (e.g., trust, networks, norms of reciprocity) in watershed communities, to the financial, technical, and human resources made available by government agencies, NGO's, academic units, and local citizens (Hardy and Koontz, 2010).

Successful collaborative watershed management programs emphasize active stakeholder engagement, employ integrated solutions, recognize the authority of multiple agencies and jurisdictions, and build on expertise and resources across sectors. Out of bio-geophysical necessity, managing a watershed involves coordinated stewardship of the waterbody and the land area that the waterbody drains. Consequently, watershed conservation and rehabilitation is typically a function of an array of public and private programs. Representatives of local, state, and federal agencies; nonprofit group; and for-profit businesses each must bring complementary resources to the task (Golden, 1998). Ideally, collaborative watershed management refers to shared decision-making and implementation by public and private sector partners who share the common goal of conserving or enhancing hydrologic resources (Michaels, 2001).

The collaborative watershed approach is on the agenda of the federal government in the United States. President Clinton's 1998 Clean Water Action Plan explicitly promotes such an approach nation wide. The plan encourages states to work with watershed stakeholders, including interested citizens, to identify watersheds with critical water quality problems and to focus resources and implement strategies to solve these problems (Michaels, 2001).

By collaborating with local entities, states can facilitate ongoing learning; devise systems for measuring, monitoring, and evaluating; and disseminate best practices or model policies. They can actively engage in propagating local experiments. States have instrumental roles to play in achieving Dorf and Sabel's (1998) ideal of democratic experimentalism, where the deliberations and performance of one jurisdiction are considered in like jurisdictions. Since problems are encountered face-to-face at the local level, a critical function of the states is to build local collaborative, managerial, financial, and technical capacity (Cigler, 1998). Recognizing the political palatability of local collaborative management results in emphasizing grassroots strategies, such as community education, and local government tools, such as floodplain zoning (Blatt, 1993).

Holistic Watershed Management

Embraces the idea that all aspects of the watershed human resources, economic development, environmental quality, infrastructure development and public safety must be considered in a holistic watershed management decision-making process. Holistic watershed management's fundament approach is in a facilitated process designed for the integration of organizations and individuals having environmental knowledge, skills and resources in the water quality and comprehensive community planning.

Consider the following roles agencies could play in sustainable holistic watershed management decision-making:

- Catalyst–incentives or regulation enforcement to improve watershed environment (Water Quality). Agency representative living in the watershed experiencing a problem.
- Responsive/Supportive–provide technical resources as needed for sound holistic watershed management decision–making.
- Stand back and let local people control the holistic watershed management planning process (UOEWQP, 2004).

Integrated Watershed Management

Previous water management efforts that were sectoral, technological and centralized have proved inadequate, because they failed to recognize and appreciate the intricacies and inter-relations of ecosystems (Pereira, 1973). Consequently, integrated watershed Management has been suggested as a solution and has been tried for decades in several countries in the world (Bowden, 1999; Mitchell, 1990; Bulkley, 1995; Lant, 1999; Pereira, 1973).

An "integrated watershed management" approach should strive to create settings for collaboration and innovation by facilitating dialogue among local stakeholders. The over-riding charge under the piloting of this approach is fostering a framework for dialogue among stakeholders for problem solving examining interdisciplinary solutions that are inherently multi-objective. That is, solutions able to address more than one problem simultaneously while addressing the entire resource based on local circumstances. The Integrated Watershed Management Program proposes a framework for fostering interdisciplinary on-ground implementation activities. Interdisciplinary takes on a meaning of multiple dimensions and scales. In one instance vertical dimensions: encompassing both surface water and ground water quality at the watershed scale. In the other instance, the lateral dimension considering the varied land uses and land covers associated with agriculture, silviculture, mining, and hydrologic/habitat modification activities, as well as those associated with urbanization (e.g., land development, transportation, recreation, etc.). These land uses and activities give rise to varying degrees of non-point source pollution or polluted runoff, which is the major contributor to impaired waters (National Research Council, 1999).

Over the past two decades, there have been numerous applications of integrated watershed management worldwide. For example, integrated watershed management approaches have been recently used for combating drought in the Jhabua watershed in India (Singh et al., 2002), assessing and managing water resources in the upper Chao Phraya in Thailand (Padma et al., 2001), assessing and managing agricultural phosphorus pollution on the Chesapeake Bay (Sharpley, 2000), tackling the problem of land degradation in Australia (Ewing, 1999), and managing the Truckee River in Nevada (Cobourn, 1999). Also, in the United States, the USEPA has been quite instrumental in promoting the integrated watershed approach to management (National Research Council, 1999).

The lessons learned from these and other initiatives indicate that in order to succeed, integrated watershed management must be participatory, adaptive and experimental, integrating all the relevant scientific knowledge/data and user-supplied information regarding the social, economic and environmental processes affecting natural resources at the watershed level (Steiguer et al., 2003). This is due to poor integration and coordination, which is either fostered or hindered by a complex set of environmental and socio-economic and institutional factors at various spatial levels such as "(1) legislation and regulations, (2) policies and guidelines, (3) administrative structures, (4) economic and financial arrangements, (5) political structures and processes, (6) historical and traditional customs and values and (7) key participants or actors" (Mitchell, 1990).

ORGANIZATIONAL APPROACHES IN WATERSHED MANAGEMENT

With the increase in population water has became a key resources. The result of the development efforts of the countries, demand on the water has been become on maximum level. This situation thread clear water resources and watersheds areas were under the pressure of human activities. So that, many of the international organizations have considered on this problem. In addition to this, they have produced many protection projects to solve water issues. In this study, we compile and evaluate the state of the art, with a special focus on material, which is produced by international organizations such as the World Bank, IFAD, FAO, IWMI.

World Bank

Over the past seven years since the World Bank's Water Resource Strategy was created, the Bank's water lending commitments have increased significantly, the quality of the water portfolio was turned around, and the outcome project rating now even outperforms the Bank average. The 2010 mid-cycle implementation progress report, "Water for All in a Changing Climate", and "The Sustainable Infrastructure Action Plan" continue to guide the Bank's strategy for managing water resources. The World Bank increasingly works with partners from multilateral and bilateral agencies, but also agencies within civil society and the private sector at the global, regional, and country levels. It's partners are Global Water Partnership (GWP), International Commission on Irrigation and Drainage (ICID), International Water Association (IWA), UN-WATER, Water Supply and Sanitation Collaborative Council (WSSCC), World Water Council (WWC).

Watershed Management Approach Before the 2000

The watershed management approach became prominent in developing countries in the 1970 in programs designed to improve upland natural resource management in order to protect downstream resources and infrastructure, when the problems of watershed degradation first became apparent.

The origins of modern watershed management can be traced to several parallel and independent movements: the restoration of the Alps, the conservation movement in the United States in the 1930s; and the watershed rehabilitation activities of colonial governments in Africa (Dargouth et. al, 2008). National and regional programs were set up to address the damage to downstream infrastructure caused by degradation in the uplands. For instance, Indonesia created a National Watershed Development Program in 1976. From the mid-1970s, Brazil launched soil conservation programs that evolved by the mid-1980s into the Integrated Soil and Water Management Program in Microcatchments. India created the National Watershed Development Program for Rainfed Areas in 1990 (Dargouth et. al, 2008).

The watershed management projects in the 1970s and 1980s consisted on specific on-site and downstream physical outcomes, this works have been done only to protect resources by the engineer approaches. So investments were high cost and not always well justified, and the assets and benefits created often had a limited life (Dargouth et. al, 2008). And ecological dimension has never seen in these works. Because of these approaches, there is no chance to put a sustainable management approach in a holistic way.

First generation of watershed management projects in the 1970s and 1980s were to prevent rapid runoff of water, slow down siltation of reservoirs, and limit the incidence of potentially damaging flash flooding. Targets were fixed in relation to physical outputs rather than economic and natural resource outcomes and a top-down planning approach was generally adopted (Dargouth et. al, 2008). In pratical way local people could not adapted to this planning outputs. So, general result of this top-down planning approach was a total mass on side. Nowadays, local users empty these old infrastructure buildings, which have been built by this plan and authors and scholars try to remediate this areas. The need to improve the livelihoods of the poor upland population was not completely ignored, but the technical improvements for agricultural production introduced usually focused on costly vegetative and mechanized technologies (Dargouth et. al, 2008). Watershed planning was based on land capability rather than on the capacities and needs of local people who lived there. People were hired as manual labor, and projects provided subsidies to stakeholders as incentives to participate (Dargouth et. al, 2008). There was little involvement of the communities in planning and implementation, which often resulted in weak commitment to the project. Benefits were expected to trickle down to the local population, although this generally happened only on a limited scale. It is given some information about water-sheet management studies belong to different years in the table 1.

Table 1. Timeline of watershed management practices (Dargouth et. al, 2008; Chambers et.al. 1989)

	Main Purpose	**Applications**	**Results**
1970s	-To protect on-site and downstream resources and infrastructure of site	-physical measures designed to manage soil and water resources in medium or large river valleys -to stabilize uplands through erosion control infrastructure	High government staff turnover and poor supervision resulted in inconsistent project management and unaccountable implementation arrangements. Projects often ignored crucial inter-sectoral linkages, resulting in lack of collaboration and communication across sectors.

1980s	-Beyond the engineering approaches acceptance of large scale collaboration between national and international organizations	-To protect to soil, water and vegetation resources -To try to decrease of local community poverty -Ecological and institutional sustainability of the resources management -To try watershed management approaches in dry land areas	-the comparative failure of the "engineering-led" approach was clear -lack of collaboration and communication across sectors
1990s	-To resource, use productivity, livelihood improvements, and poverty reduction objectives in addition to resource conservation.	-To identify and execute interventions within an integrated farming systems approach, including lowland and upland agriculture, pasture, and forest management- To adopt a participatory and demand-driven development approach, which was influenced by emerging theories of "farmer first"	-To led watershed management programs to seek ways to build on existing social structures and institutions and to support planning and implementation by community institutions the promotion of low-cost vegetative techniques for erosion control replaced or complemented the previous "mechanical" techniques that had relied on heavy construction and had proved expensive to build and difficult to maintain

A number of countries as diverse as India, Kenya, and Nigeria have seen environmental rehabilitation carried out almost spontaneously as population pressure increased or environmental degradation, particularly erosion, reached high levels. For example, Tiffen, Mortimore, and Gichuki (1994) explored the relationship between increasing population density, productivity, and environmental degradation through a case study in southeast Kenya. They showed that population increase combined with market opportunities stimulated local investment and innovation in dry land farming and environmental recovery of heavily degraded land (Dargouth et. al, 2008).

World Bank Projects

There are many water management projects which were held by World Bank on the issue that the Participatory Watershed Management. In the same time they want to support local economies in the same time by the way, they want to protect natural sources. These projects may include financing critical economic infrastructure such as rural footpaths, sewerage, drainage, roads and water supply.

The development objective of the proposed Participatory Watershed Management Project (PWMP) is to improve the productive potential of natural resources and increase incomes of the rural households in selected watersheds in Jammu & Kashmir using socially inclusive, institutionally and environmentally sustainable approaches. A secondary objective is to support policy and institutional development in in Jammu & Kashmir to harmonize watershed development projects and programs across the state in accordance with best practices (Darghouth et. al., 2008).

Most of the key lessons stem from the previous project, although they are hallmarks of other successful watershed development projects in India (Darghouth et. al., 2008):

- Cost-sharing is critical for ensuring ownership and continued operations of project-financed assets. Hand-outs and full subsidies are to be discouraged.
- Transparency of decision-making and resource allocation is important to ensure that all community members are fully aware of project activities and benefits. Several participatory planning tools,

such as social mapping and village account displayed in central places, have been used successfully in this respect.

- As a land-based approach, the watershed development project needs to take into account marginal groups (women, landless, nomadic tribes) who may not directly benefit from project interventions unless targeted efforts are made by the project to cater to their unique needs.

- The proposed project should leverage its resources going beyond target watersheds and encouraging state-wise initiatives in capacity-building and policy reforms that will yield wider benefits (Darghouth et. al., 2008).

National policies on watershed management have generally been driven by program experience rather than vice versa.

Initially watershed management programs were adopted in the 1970s and 1980s by governments as pragmatic responses to natural resource degradation and the related social and economic costs. These programs developed in an iterative fashion, with early setbacks over approaches dominated by engineering being succeeded by tests of community-based approaches targeting sustainable changes in land use practices. Now, in several countries success in testing community-based approaches has led to adoption of broader policies for community-based watershed management (Darghouth et. al., 2008).

Turkey has built on local and regional experience to formulate policies for community based watershed management in poor upland areas. The Turkish rural economy has been characterized by a high incidence of poverty, particularly in upland areas. The consequent growing pressures on forests and pasture have reduced vegetative cover and diminished soil fertility and the carrying capacity of rangeland. This has contributed to reductions in infiltration rates and to increases in peak river flows, flooding, and sedimentation problems (Darghouth et. al., 2008).

Beginning in the late 1980s, Turkey tested an integrated and participatory approach to watershed management in a number of micro-watersheds, and from 2004 expanded the approach to three major river basins. Policy is now based on a community-driven approach to natural resource management, integrating forestry, soil and water conservation, and crop and livestock production. The government shares the cost of a mutually reinforcing package of resource use productivity enhancing and conservation measures. This policy has driven institutional change, particularly the coordination and integration of the activities of different government departments at the microwatershed level and the development of watershed-based forest resource management plans (Darghouth et. al. 2008).

There are some projects which has been held by World Bank in Turkey where the two continent are joined. One example of this projects is Eastern Anatolia Watershed Project. This subproject would seek to identify and establish in-situ conservation areas in Turkey, for the protection of genetic resources and wild relatives of important crops and forest tree species that originated in Turkey. It will provide for sustainable in-situ conservation of genetic resources in cereals, horticultural crops, medicinal plants, forest trees, and pasture grasses and legumes through an integrated ecosystem approach. It will also contribute to the broader objective of conserving sustainable farming and forest systems that is a key element of the Watershed Rehabilitation project. The project has been developed the institutional capacity in Turkey for preparing and implementing a national strategy for in-situ conservation which could include landraces in a second phase Project. The aim was to test and develop a new approach to conservation of genetic diversity which has not been tried on a large scale anywhere in the world (Anonymous, 2012c).

Dams influence both water quality and the very functionality of rivers, of planetary life cycle processes harming people, culture and nature, collectively. Roughly two-thirds of the world's rivers have suffered harm from the ten thousands of dams that have been built over the past century. Many of the world's great rivers such as the Indus, the Colorado, and the Yellow Rivers, no longer reach the ocean, turning once-productive deltas into biological deserts. More than tropical rainforests, marine environments, or coastal wetlands, freshwater ecosystems are experiencing the greatest loss of biodiversity, in large measure due to dams. Over the past 40 years, freshwater ecosystems have lost 50% of their populations and over a third of remaining freshwater fish species are threatened with extinction. Presently, the great river basins of the world are experiencing a new wave of damming: The Amazon, the Mesopotamia, the Congo, the Mekong are each superlative in their contributions to planetary cycles, biodiversity and cultural evolution of human civilization. Each of these basins is threatened with audacious and narrow-sighted schemes that will irreversibly disconnect rivers and cost the planet billions in lost ecosystem services. The main reason underpinning this new wave is that dams are misinterpreted as "renewable energy"and promoted as solution to climate change. Yet, as proven different parts of the world, such as Amazonia and Mesopotamia, dams do not generate renewable energy, but irreversible natural and cultural destruction. Also scientific studies indicate that dams and reservoirs are globally significant sources of the greenhouse gases carbon dioxide and, in particular, methane (Anonymous, 2012d).

Dams have significant adverse effects on heritage through the loss of local cultural and archaeological sites and resources. For instance, Ilisu Dam will destroy more than 300 archaeological sites, including the 11,000 year-old historic town of Hasankeyf that is the only place in the world that meets nine out of 10 criteria for UNESCO World Heritage Site status (Anonymous, 2012d).

Dams also cause loss or damage of cultural heritage through land reclamation and irrigation projects and the construction of power lines, roads, railways and workers towns. In most cases, no measures have been taken to minimize or mitigate the loss of cultural and archaeological resources (Anonymous, 2012d).

Figure 1. Hasankeyf, a demonstration against the construction of the Ilisu Dam, held by Turkish Nature Organization and general view of the area (Anonymous, 2012e and Anonymous, 2012f).

International Fund for Agricultural Development (IFAD)

International Fund for Agricultural Development (IFAD), a specialized agency of the United Nations, was established as an international financial institution in 1977 as one of the major outcomes of the 1974 World Food Conference. IFAD is dedicated to eradicating rural poverty in developing countries (Anonymous, 2012g).

In recent decades, IFAD has supported changes in land and water governance as a way to improve rural poor people's access to these natural resources, and to ensure poverty reduction, increased food security and better livelihoods. This involves working through community-based and civil society organizations and NGOs to better identify the changes that are needed, and with national and local governments to change policies and legislation. The aim is to empower rural poor people to participate in managing the common property resources on which they depend.

Paid environmental or watershed services are increasingly recognized as a potential source of additional income for poor rural people. IFAD's Environmental and Social Assessment Procedures, which include operational statements, for example on irrigation, range resources, inland fisheries and wetlands, regulate exploitation of water resources and the environment.

The State of the World's Land and Water Resources for Food and Agriculture (SOLAW) analyses a variety of options for overcoming constraints and improving resource management in these areas of heightened risk. In each location, a mix of changes in institutional and policy measures will have to be combined with greater access to technologies for better management of land and water resources. Increased investments; access to novel financing mechanisms; and international cooperation and development assistance will also help overcome these constraints (Anonymous, 2012g).

This first issue of SOLAW, which complements other "State of the world" reports published regularly by FAO, is intended to inform public debate and policy-making at national and international levels. IWMI is one of 15 international research centers supported by the network of 60 governments, private foundations and international and regional organizations collectively known as the Consultative Group on International Agricultural Research (CGIAR). IWMI's Mission is to improve the management of land and water resources for food, livelihoods and the environment. IWMI targets water and land management challenges faced by poor communities in the developing world/or in developing countries and through this contributes towards the achievement of the UN Millennium Development Goals (MDGs) of reducing poverty, hunger and maintaining a sustainable environment. These are also the goals of the CGIAR (Anonymous, 2012g).

In most developing countries, agriculture accounts for more than 80 per cent of water use. An increased focus on this sector is needed to address the water crisis. For poor smallholder farmers, water and land cannot be treated as separate issues. Government and development actions targeting only land or only water governance changes are unlikely to achieve sustainable impact. For poor farmers, secure land access can lead to secure water access, and the reverse is also true. This in turn leads to access to credit and investment in their farms, which can improve their livelihoods, and improve agricultural water use efficiency (Anonymous, 2012g).

At local level, land and water governance structures already exist in some form, but these are not systematically recognized at higher institutional levels. If reforms are to improve the livelihoods of rural poor people, their voices and concerns need to be heard and acknowledged as part of the reform process.

Indigenous knowledge and practices need to be recognized as a strong basis for building lasting change in land and water governance (Anonymous, 2012g).

The role of women in land and water management and use must be recognized as part of the reform process.

The capacity of individuals, communities and NGOs must be developed so they can take on the responsibilities associated with reforms. Building trust in communities and among partners is an essential part of capacity development, so that they can act collectively for mutual benefit. Sufficient time must be allowed for enlisting broad support for reform. NGOs can be a useful vehicle to support the reform process, which does not always fit with time and budget constraints of development projects. External support is important. Financial support, combined with policy dialogue, can be a catalyst for NGOs, communities and governments to pursue change. International agencies also provide small communities and local and national organizations with a valuable sense of international recognition, credit and encouragement (Anonymous, 2012g).

Documenting land and water governance experiences in a country promotes better understanding of stakeholders' and IFAD partners' views. This creates opportunities for effective pro-poor advocacy and policy dialogue at all levels aimed at improved access to productive natural resources and technology (Anonymous, 2012g).

Food and Agriculture Organization of the United Nations (FAO)

As a knowledge organization, FAO creates and shares critical information about food, agriculture and natural resources in the form of global public goods. However, this is not a one-way flow. FAO plays a connector role, through identifying and working with different partners with established expertise, and facilitating a dialogue between those who have the knowledge and those who need it. By turning knowledge into action, FAO links the field to national, regional and global initiatives in a mutually reinforcing cycle (Anonymous, 2012h).

In assessing the anticipated impacts of climate change on agriculture and agricultural water management, it is clear that water availability (from rainfall, watercourses and aquifers) will be a critical factor. Substantial adaptation will be needed to ensure adequate supply and efficient utilization of what will, in many instances, be a declining resource (Anonymous, 2012h). By the definitions of the FAO; the world's food production depends on the availability of water, a precious but finite resource and nearly half of the world's population is affected in various ways by mountain ecology and the degradation of watershed areas.

International Water Management Institute (IWMI)

IWMI is one of 15 international research centers supported by the network of 60 governments, private foundations and international and regional organizations collectively known as the Consultative Group on International Agricultural Research (CGIAR). It is a non-profit organization with a staff of 350 and offices in over 10 countries across Asia and Africa and Headquarters in Colombo, Sri Lanka (Anonymous, 2012ı). IWMI›s Mission is to improve the management of land and water resources for food, livelihoods and the environment. IWMI›s Vision, reflected in the Strategic Plan is water for a food-secure world. IWMI targets water and land management challenges faced by poor communities in the developing world/or in developing countries and through this contributes towards the achievement of the UN Millennium Development Goals (MDGs) of reducing poverty, hunger and maintaining a sustainable environment. These are also the goals of the CGIAR (Anonymous, 2012ı).

Research is the core activity of IWMI. The research agenda is organized around four priority Themes including Water Availability and Access; Productive Water Use; Water Quality, Health and Environment; and Water and Society. Cross cutting activities in all themes include, assessment of land and water productivity and their relationship to poverty, identification of interventions that improve productivity as well as access to and sustainability of natural resources, assessment of the impacts of interventions on productivity, livelihoods, health and environmental sustainability. IWMI works through collaborative research with many partners in the North and South and targets policy makers, development agencies, individual farmers and private sector organizations. IWMI's new triple approach to uptake is being implemented (Anonymous, 2012ı). This involves:

- Project uptake strategies: These are targeted strategies built into projects at the beginning. They are focused on the project results and the potential users of the results.
- Regional uptake strategies: These are particularly important to keep the momentum going when projects are completed, and to build linkages across projects to provide synthesized messages.
- Macro uptake strategies: This involves making all information and knowledge available as broadly as possible, making it easily accessible and promoted widely (Anonymous, 2012ı).

IWMI supported to some project about ecosystems for water and food security. The result of water management in agroecological landscapes is remarkable.

Water Management in Agroecological Landscapes

Managing agricultural land to deliver multiple services considerably improves values derived. This is best done at the landscape level, linking ecosystems and managing natural resources such as water and land specifically to enhance ecosystem services, thereby integrating all sources of water in the basin, from rain, in the soil, in aquifers and as surface water. Water is then no longer supplied to crops, trees, livestock, or fish, but to multifunctional agroecosystem linked and managed together at the river basin or landscape level. In this way, synergies can be exploited and productivity can be improved, while obtaining benefit from improved carbon storage, erosion control, water retention, waste treatment, and cultural and recreational values including tourism. Notably, most of these added services do not conflict with agricultural production but in many cases improve both its productivity and sustainability.

The recommendations are based on the findings on the various types of ecosystems and integrated ways to enhance water management for food security. In addition, principles of sustainable water management for agriculture in ecoagriculture have been incorporated (Molden et. al. 2007). As a result, the specific recommendations concerning water are only part of the larger message on how to manage natural resources sustainably and make the transition from production systems to multipurpose agroecosystems.

These guiding principles have been combined from various sources and would increase ecosystem services and sustainability while using the same resources, hence be more productive (Swift et al., 2004; Molden et al., 2007; Van der Zijpp et al., 2007; Bossio and Geheb 2008; Hajjar et al., 2008;Swallow and Meinzen-Dick 2009; World Bank 2009; Zomer et al., 2009; Garrity et al., 2010;McCartney and Smakhtin 2010):

Promote diversity within the production systems: Optimizing the diversity of the above and below ground biotic components within the production system (crop biodiversity, animal diversity, soil biodiversity, and pollinators) can increase the adaptive capacity of the cropping system to buffer against fluctuations in water availability, temperatures, pests and diseases, thereby enhancing the resilience of rural livelihoods. Synergies with livestock and aquaculture can be explored to increase resource recovery and productivity, for instance in crop livestock systems, rice-fish culture, tree-crop systems, aquaculture in reservoirs, forest-pastures, or wastewater-fed aquaculture. Integration of trees can help fix nitrogen, tighten nutrient, water, and carbon cycles, and produce additional goods, such as year-round availability of fodder and biomass for use as organic fertilizer and fuel.

Promote diversity in landscapes: Landscapes with high levels of biodiversity are more resilient and better able to mitigate environmental impacts. Large mono-cropped areas can be developed into landscapes with higher levels of biodiversity by identifying and linking natural habitat patches, including aquatic ecosystems. Habitat integrity and connectivity can be maintained by incorporating hedgerows, multipurpose trees, and corridors of natural vegetation interconnecting parcels of agricultural land and natural ecosystems (such as wetlands and forests – these may need to be specifically developed where they are too far away). In large irrigated areas, canals and roads can be lined with perennial vegetation such as trees, thereby also serving as important passages and habitats for animals. Canals and other waterways can connect aquatic ecosystems and thereby maintain the connectivity of migratory routes, providing the variety in habitats required for subsequent life cycle stages such as spawning. Landscape-scale planning of strategic tree cover interventions can reduce flow accumulation by providing sites for water infiltration and penetration. By incorporating both fodder production and grazing land, livestock can be managed at the landscape level too, thereby enabling animals to reach otherwise inaccessible feed sources and avoiding overgrazing and trampling of vulnerable areas.

WATERSHED MANAGEMENT STUDIES IN DEVELOPING COUNTRY

For several decades, Integrated Watershed Management has been suggested and tried in several countries in the world, as an effective way to address complex water resource challenges. However its implementation has not been successful in most cases, due to various barriers. In Kenya, this approach is new and requires appropriate strategies to overcome these barriers and stimulate effective integrated watershed management. To design suitable and effective strategies, there is need to understand institutional features at various spatial levels, which promote or hinder integration and coordination. Wamalma's paper therefore explores the prospects and barriers of integrated watershed management of Mara, by examining the existing complex set of biophysical and socio-economic conditions, stakeholders attitudes and perceptions, arrangements for participation of stakeholders, available institutional structures and financial plans, and recent policy reforms in water and forestry sectors. Empirical information was gathered from official documents, direct observations, semi-structured interviews with managers, administrators, politicians and households of Mara watershed. Results indicates that, integrated watershed management of Mara is likely to be fostered by the critical biophysical and socioeconomic conditions, suitable institutional structures that are being established, water and forestry reforms, recognition of stakeholder participation and enhanced education of stakeholders, and leverage of resources from NGOs. However this efforts are likely to be hampered by disparity of views and perspectives, egoistic tendencies of influential leaders, inadequate financial plans, lack of effective coordination mechanisms, ineffective multi-stakeholder process, unwillingness of the local community and the politicians if there interests are not addressed, and lack of legitimacy for the institutional structures that are being created. This study therefore suggests adoption of multi-stakeholder forums for building understanding and bridging the disparity of views, the need to address the legitimate interests of the local community and politicians, and enhanced level of understanding among stakeholders on the interactions and interdependencies among the variables of the ecosystem. Finally the study recommends the need for a more effective coordination arrangement such as elevation of the CAAC (Catchment Area Advisory Committee) to the level of a coordinating agency instead of relying on a single government agency like WRMA (Water Resources Management Authority) as provided for by the water act 2002 at the catchment level. In conclusion as much as the new water sector reforms provide legitimacy, impetus and a framework for integrated watershed management in Mara and other water catchments in Kenya, implementation may still remain a challenge if barriers are not identified and addressed.

Liu et al.'s study highlights two highly degraded watersheds in the semi-arid Amhara region of Ethiopia where integrated water resource management activities were carried out to decrease dependence on food aid through improved management of 'green' water. Whiletop-down approaches require precise and centrally available knowledge to deal with the uncertainty in engineering design of watershed management projects, bottom-up approaches can succeed without such information by making extensive use of stakeholder knowledge. This approach works best in conjunction with the development of leadership confidence within local communities. These communities typically face a number of problems, most notably poverty, that prevent them from fully investing in the protection of their natural resources, so an integrated management system is needed to suitably address the interrelated problems. Many different implementing agencies were brought together in the two study watersheds to address water scarcity, crop production, and soil erosion, but the corner stone was enabling local potential through the creation and strengthening of community watershed management organizations. Leadership training and the reinforcement of stakeholder feedback as a fundamental activity led to increased ownership and willingness to take on new responsibilities. A series of small short term successes ranging from micro-enterprise cooperatives to gully rehabilitation have resulted in the pilot communities becoming confident of their own capabilities and proud to share their successes and knowledge with other communities struggling with natural resource degradation (Liu et al., 2008).

Specifically it has learned that (1) communities have high interest in development initiatives, including sustaining natural resources; (2) only when people make real contributions of their own resources will they strive to ensure the implementation of the planned activities; (3) emphasis must be given to effective organization of communities rather than only focusing on technology development; (4) including non-technical leaders in the information loop provides great benefits at the community level. Introduced ideas, such as the fair representation of women and poor in the management process, women's micro-enterprise groups, and closing off areas to livestock, faced strong cultural opposition at the beginning but have proven very successful thanks to trial tests and continue collaboration efforts by farmers, non-technical leaders, and project personnel. It has also found that communities most at risk were also the most open to new ideas (Liu et al., 2008). Partnerships and co-operative environmental management are increasing world wide as is the call for scientific input in the public process of ecosystem management. In Hawaii, private land owners, non-governmental organizations, and state and federal agencies have formed watershed partnerships to conserve and better manage upland forested watersheds. In this paper, findings of an international workshop convened in Hawaii to explore the strength of approaches used to assess take holder values of environmental resources and foster consensus in the public process of ecosystem management are presented. Authors draw upon field experience in projects throughout Hawaii, Southeast Asia, Africa and the US main land to derive a set of lessons learned that can be applied to Hawaiian and other watershed partnerships in an effort to promote consensus and sustainable ecosystem management. Interdisciplinary science-based models can serve as effective tools to identify areas of potential consensus in the process of ecosystem management. Effective integration of scientific input in co-operative ecosystem management depends on the role of science, the stakeholders and decision-makers involved, and the common language utilized to compare tradeoffs. Trust is essential to consensus building and the integration of scientific input must be transparent and inclusive of public feedback. Consideration of all relevant stakeholders and the actual benefits and costs of management activities to each stakeholder is essential. Perceptions and intuitive responses of people can be as influential as analytical processes in decision-making and must be addressed. Deliberative, dynamic and iterative decision-making processes all influence the level of stakeholder achievement of consensus. In Hawaii, application of lessons learned can promote more informed and democratic decision processes, quality scientific analysis that is relevant, and legitimacy and public acceptance of ecosystem management (Gutrich et. al. 2005).

Maintaining base-flows in the Pocono Creek is a particular concern. Knowing that the demand for water increases as competition for existing water resources grows with a growing number of users, concerns about sustaining the exceptional water quality and supply for future generations were raised throughout the 2003 Pocono Creek Pilot Project for Goal-based Watershed Planning (Pilot), which was funded by a PA Growing Greener grant. Local resource managers and citizens emphasized this issue during both the Pilot's goal setting (2000) and project evaluation stages (2003). The Environmental

Protection Agency's Region 3 Office (EPA-3) organized and funded the current project, known as the Framework for Sustainable Watershed Management (Combe et. al. 2009).

The purpose of the Framework is to introduce a program that balances growth with natural resource protection, so that future generations can enjoy the highly valued natural resources of the region, while enjoying economic prosperity. This program is accomplished in three stages; technical, planning and watershed community outreach. The technical stage identifies the impacts of rapid growth on a watershed's water resources. The planning stage develops management strategies that balance regional growth needs with natural resource protection. The planning and watershed community outreach effort introduces an innovative approach to protecting the region's water resources through a community-wide public art event that receives extensive media coverage, has high visibility and generates enthusiastic community participation (Combe et. al. 2009).

The micro-watershed has proved a flexible and practical unit for project implementation and has reduced costs. However, the definition of a micro-watershed needs to be adapted to the social, administrative, and physical context. Best practice is that choice of scale should be driven by a participatory analysis of problems throughout the watershed, preferably within a broader watershed planning framework, as was done in the Loess II Project in China. In 1994 the World Bank funded one of the most successful conservation projects in the world, which improved the local environment, but also boosted the livelihoods of more than 1 million farmers. The number of people living in poverty in the region dropped from 59 percent in 1993 to 27 percent in 2001 and almost 3,000 square kilometers of eroded land was terraced (World Bank, 2003).

The restoration project worked closely with local farmers to build terraces, plant shrubs and trees, install irrigation and sediment control dams, and provide training and education on erosion control.

This watershed restoration project is now being replicated across China, with continued success (World Bank, 2003).

Based on this, programs can be clear from the beginning about the proposed scale of interventions and the socioeconomic, environmental, and technical criteria for defining the micro-catchment and for selecting which micro-catchments to target. The micro-watershed approach also raises some difficulties when it comes to scaling up. Working at the micro-watershed scale does not necessarily aggregate or capture upstream-downstream interactions. A patchwork of upstream interventions would only have a significant impact downstream if prioritized and planned within the larger watershed context and with understanding of the spatial and hydrological links between the perceived externalities and their causal factors (for example, land and water use). The lesson is that integration of watershed management activities beyond the micro-watershed requires higher level technical planning. In best-practice approaches, planning includes an institutional mechanism where stakeholders have a voice and are able to agree on measures from the micro-watershed scale upwards that can achieve both local and larger-scale objectives. The approach also needs to deal with institutional challenges of interagency collaboration and local-regional–level coordination.

Groetschel's (2000) study has investigated the problem situation in selected villages in Kachchh and Dahod district and looked into the institutional framework of watershed development in Gujarat. Field surveys for the target area analysis were carried out in six villages. In the five priority districts for watershed development activities, 15 NGOs and a number of Government institutions were met and assessed. The study is part of the planning process for the Indo-German Watershed Development Program (IGWDP) in Gujarat. It is geared towards providing a qualitative description of important issues for a region specific watershed development approach and gives recommendations for possible adaptations of the watershed development approach. In answer to the problem of natural resource degradation, more than 1200 watershed development projects have been implemented under different programs by the Rural Development Department in Gujarat since 1995. More than 70 percent of these are operated by NGOs. All government funded watershed development projects in Gujarat follow common guidelines, which determine implementation strategies, program content and components, principles of project management, capacity building, financial aspects and monitoring and evaluation. Major aspects of the approach include sustainability, participation, empowerment and decentralization.

In general, the spirit of the common guidelines must be considered to be appropriate. However, as nation wide guidelines, they lack considerations of regional characteristics and problems. More flexibility would be required in many aspects to ensure appropriate handling of local problems. As an integrated, but, basically land based approach, watershed development needs careful consideration of equity concerns. Soil and water conservation measures alone might otherwise further benefit the rich instead of fostering social and economic cohesion. Implementation problems also arise from cooperation difficulties among different GOs and NGOs. The treatment of forest lands and common property resources, as required in most cases when following the ridge to valley approach, faces many difficulties. For the last years, the emerging Joint Forest Management Program (JFM) has been trying to mitigate some of those problems. Meanwhile it is the largest program of the Forestry Department. In addition to JFM, there are other government programs, which are supplementary to and supportive of watershed development efforts. The status of NGOs in Gujarat is very strong. Many of them have been involved in watershed development related activities for a number of years already. They actively participate in policy dialogues with the Government and are a driving force in pursuing adaptation of the watershed development approach. Many of the NGOs have developed special implementation strategies, often depending on their specific field of interest or the background of their organization and staff.

Other NGOs are implementing regionally adapted solutions based, for example, on the problem of salinity in coastal areas. In 1999, the National Bank for Agriculture and Rural Development (NABARD) joined the watershed development efforts by establishing its Watershed Development Fund. The fund aims at further strengthening participatory watershed development initiatives. The selection criteria for watersheds are a significant proportion of Scheduled Castes/Scheduled Tribes (SC/ST) population, high extent of rainfed farming and a high potential for watershed development. The regional watershed management cell of NABARD, which took up work in August 1999, is planning to undertake 19 projects at the first stage. Number and staff composition will have to widen, when the program expands.

Successful project implementation demands a range of skills and attitudes at village, NGO and program management level, which are not always or not sufficiently developed. Capacity building, therefore, plays a central role. Requirements at village level include raising awareness for environmental problems and resource management. Additional training will be required in order to make best sustainable use of conserved resources. Many NGOs in Gujarat have been successfully involved in watershed development projects. They have gained ample experience with this approach. What is still needed in some cases is to balance any existing bias in their work, which in many cases is either on the social or on the technical side. However, capacity building for the proposed IGWDP in these respects will probably not require the establishment of a separate institution. There are well established networks, which take care also of capacity building requirements. Requirements at the management level will mainly comprise comprehensive supervisory and monitoring functions.

An important goal of watershed management is to ensure the sustainable development of water resources. However, since integrated watershed management is much easier said than done, this approach is going to illustrate a real watershed, the Big Lost River watershed located in south-eastern Idaho, that has suffered from practical problems and examine this watershed in terms of sustainability just before and after the implementation of this approach. The Big Lost River watershed is about 3730 km^2 and drains into the Big Lost River and its tributaries, which in turn drain to the Eastern Snake River Plain. The Big Lost River is one of the largest tributaries to the Eastern Snake River Plain. Although the majority of the land in the Big Lost watershed are forest and grass land, agriculture, animal raising, and residential development make significant contributions to the water quality and quantity problems.

To achieve the goals of both the stakeholders and the management decision-makers under sustainability conditions, it is necessary to implement the following steps:

- The existing characteristics of the Big Lost River watershed should be assessed. This will (a) include determining problems that can hinder the watershed management effort, (b) classifying these problems into groups (scientific, social, or motivational), and (c) determining the priorities in solving these problems.
- Suitable solutions to solve these problems should then be proposed such as:
- developing useful relationships among selected water quality parameters to make evaluation of the water quality situations easier by using a small number of parameters,
- improving integrated management of surface and ground waters by estimating uncertainty in the water budgets,

- exploring interactive methods for effective public participation, and
- determining the impact of water quality deterioration and the increase in water demand on water-use physical sustainability.
- These solutions should be evaluated by examining the sustainability. This approach seeks to introduce a pertinent watershed management framework that can be used to bring Together major stakeholders in a watershed in an-unified effort to protect and improve the quality and quantity of their watershed. The approach can be shared among decision-makers and stakeholders and the results can be used to evaluate and achieve water quality improvements and water quantity conservation. The study tackles watershed problems using a robust technique based on watershed management feasibilities. While this technique has scientific tools to solve watershed problems, it also uses sustainability and public participation as great issues of socio-economic values. The examination of water quality effects, especially erosion of lands and increasing sedimentation as well as increases in water demand due to population or land use changes, on the physical sustainability of water use in a watershed will be a significant benefit and can be applied for other watersheds. The proposed actions related to improving water quality and conserving water quantity, will lead to successful watershed management and encourage stakeholders to participate in the decision-making processes (Said et. al. 2006).

CURRENT SITUATION OF WATERSHED MANAGEMENT IN TURKEY

Watershed in Turkey

Turkey has been separated into 25 different hydrologic watershed area (Table 2. Figure 2) and annual flow of these areas are nearly 186 billion cubic meters. One of third of these flow has been calculated in the Tigris watershed area which is situated in the eastern part of the Turkey. In the spatial terms volume of these watershed has been followed by Kizilirmak and Sakarya watershed and in the term of the rain regime these watershed has been followed by the East Blacksea, East Mediterranean and Antalya Regions (Anonymous, 2013).

Social and demographic conditions and resource usage of the watershed has been changed by the aspect of the vertical and horizontal conditions of the area and it has been showed differences by the local conditions. Upper parts of the watershed and in the east regions of the country, the general population levels are commonly low levels. Rural poverty and necessities to use of natural resources to sustain life is more common in the upper part of the watershed and eastern part of the country rather than the western parts. Upper parts of the watershed main economic activities are livestock, small agricultural production and forestry. In the contrast of these situation downstream areas are widely used for agricultural products (Anonymous, 2013).

The amount of the total water resource is 112 billion cubic meter and %36 of these water is proper for use and 32 billion cubic meters are used for irrigation, 7 billion cubic meters are used for drinking water and 5 billion cubic meters are used for industrial proposes (Anonymous, 2013).

In the last decades, with the participation of the non-governmental organizations and universities, there is a great awareness in the public about the sustainable use of natural resources and its benefits (Such as soil protection, water volume and quality, carbon emission, protection of the biological diversity etc.). With the parallel these situation, there is an increase of the projects about rehabilitation works, soil protection barriers and action plans to protect natural biological diversity (Anonymous, 2013).

Table 2. The Watershed of Turkey (Anonymous, 2013)

Watershed name	**Rainfall area**		**Avarage annual flow**		**Avarage annual productivity**
	(km^2)	%	(km^3)	(%)	(l/s/km^2)
(01) Meric-Ergene Watershed	14,560	1.9	1.33	0.7	2.9
(02) Marmara Watershed	24,100	3.1	8.33	4.5	11.0

(03) Susurluk Watershed	22,399	2.9	5.43	2.9	7.2
(04) North Aegean Watershed	10,003	1.3	2.09	1.1	7.4
(05) Watershed	18,000	2.3	1.95	1.1	3.6
(06) Kucuk Menderes Watershed	6,907	0.9	1.19	0.6	5.3
(07) Buyuk Menderes Watershed	24,976	3.2	3.03	1.6	3.9
(08) West Mediterranean Watershed	20,953	2.7	8.93	4.8	12.4
(09) Antalya Watershed	19,577	2.5	11.06	5.9	24.2
(10) Burdur Lake Watershed	6,374	0.8	0.50	0.3	1.8
(11) Akarcay Watershed	7,605	1.0	0.49	0.3	1.9
(12) Sakarya Watershed	58,160	7.5	6.40	3.4	3.6
(13) West Blacksea Watershed	29,598	3.8	9.93	5.3	10.6
(14) Watershed	36,114	4.6	5.80	3.1	5.1
(15) Watershed	78,180	10.0	6.48	3.5	2.6
(16) Konya Watershed	53,850	6.9	4.52	2.4	2.5
(17) East Mediterranean Watershed	22,048	2.8	11.07	6.0	15.6
(18) Watershed	20,450	2.6	8.01	4.3	12.3
(19) Asi Watershed	7,796	1.0	1.17	0.6	3.4
(20) Watershed	21,982	2.8	7.18	3.9	10.7
(21) Fırat-Dicle Watershed	184,918	23.7	52.94	28.5	8.3
(22) East Blacksea Watershed	24,077	3.1	14.90	8.0	19.5
(23) Coruh Watershed	19,872	2.6	6.30	3.4	10.1
(24) Aras Watershed	27,548	3.5	4.63	2.5	5.3
(25) Van Lake Watershed	19,405	2.5	2.39	1.3	5.0
Total	**779,452**	**100.0**	**186.05**	**100.0**	

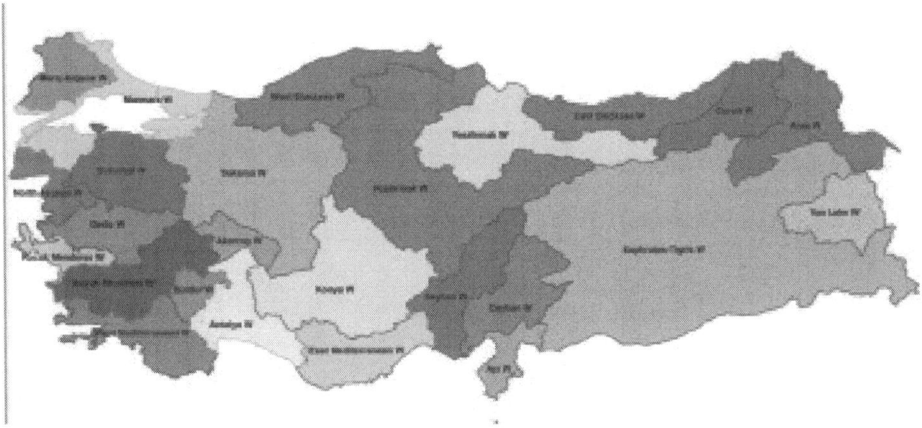

Figure 2. Watershed areas in Turkey (Anonymous, 2013).

Stakeholders About Watershed Management in Turkey

Watershed management practices in Turkey were held mainly by the governmental organizations such as ministry departments and governmental agencies. Briefly these agencies can be classified by the ministries occupation. These are include (Anonymous, 2013);

Agencies in the Ministry of Forestry and Hydraulic Works: General Directorate of Combating Desertification and Erosion, General Directorate Of Forestry, General Directorate for State Hydraulic Works, General Directorate of Water Management, Directorate of Nature Conservation and National Parks, Directorate of Meteorological Services, Presidency of Strategy Development, Turkish Institute of Water, Presidency of Information (Anonymous, 2013).

Agencies in the Ministry of Food, Agriculture and Livestock: General Directorate of Agricultural Improvement, General Directorate of Vegetative Production, General Directorate of Agricultural Researches and Politics, General Directorate of Fishery and Water productions, Department of GIS (Anonymous, 2013).

Agencies in the Ministry of Environment and Urban Planning: General Directorate of Spatial Planning, General Directorate of Control about Environmental Impact Assessment, General Directorate of Environment Management, General Directorate of Protection about Natural Heritage, General Directorate of City Bank, General Directorate of Infrastructural Works (Anonymous, 2013).

Agencies in the Ministry of Energy and Natural Resources: General Directorate of Quarry Companies (Anonymous, 2013).

Ministry of Culture and Tourism, Ministry of National Education, Ministry of Health.

Agencies in Prime Ministry: Under secretariat of Treasury, Disaster and Emergency Management Presidency

Agencies in the Ministry of Interior: General Directorate of Local Administration

Local Administrations; governorship (Anonymous, 2013).

Other Civil Stakeholders:

Non-governmental Organizations: (Organizations, groups, activists about the issues like Soil and Water Resources, Biological Diversity and Rural Development)

Profession Chambers

Watershed Groups

Rural Societies who lives in Watershed Region

Science and Education Organizations (Universities, The Scientific and Technological Research Council of Turkey, Research Institutes)

Stakeholders from Private Sector (Anonymous, 2013).

Stakeholders of Watersheds have different views and importance about economic, ecological, social and cultural by the results product (energy management, clear water resources, irrigation water, water for industrial needs, income and productivity, conservation of the biological diversity, protection of the air cleanness, recreation needs, natural landscape, eco-tourism, game sports, protection of the local cultural motivations of the watershed, national income, surplus to liberal economy) (Anonymous, 2013).

SWOT Analysis of the Watershed Management

There is a strong need to swot analysis for figuring out about the future goals and aims of the watershed management (Figure 3).

SWOT ANALYSIS	
Strengths	**Weaknesses**
Organizations have gained their experience about the watershed management and therefore they can attempt to management in a more Holistic way	Inadequacies in the terms of the watershed management politics and strategies in addition to these problems about coordination between sectorial investments
Increase in the watershed protection and monitoring action plan and river region management plans	Coordination, legality, policy problems between governmental agencies
Increase about governmental investments for the watershed organizations	Inadequacies in the term of the participation of the non-governmental organizations and local people
Increase in the terms of the soil protection and watershed rehabilitation	Problems about the sharing information between agencies about development plans
Increase in the knowledge share between governmental and non-governmental organizations	Inadequacies conditions about monitoring system which is caused by the lack of the modern investigation systems
Increase in the term of the protection area volume	Lack of national watershed monitoring data set
Biological diversity monitoring department has been established a data matrix to inventory of the biological resources	Inadequacies to set priorities about watershed projects and actions Deficiencies to set rules of the watershed plans for coordinating management actions Deficiencies in the lack about to monitoring of ecological and socio-cultural data's Deficiencies in the calculation of the cost benefit analysis about the stakeholder who effected both positive and negative Lack of the sharing information and dialog between researches and applications Lack of the stuff who should work for management agencies

Decrease of pressure on the upper watershed by stakeholders because of the mitigation	The deterioration and extinction process on limited natural resources
Access to data and possibilities of utilization of technology	The deterioration of population balance among the downstream and upper watershed
Diversity and potentials of Watershed that are caused by natural resources and living creature's	Because of increasing of the population, increase of demand and requirement from watershed areas' products and services.
Increase of awareness on the society to natural resources and environmental protection	Rural poverty
Increase of NGO's effects and contributions	Because of migration to urban, decrease of young population in the term of work force in rural area
Increase of political attention and support	Inadequacies of sensitivity and education of the watershed resources value
Development of institutional participatory approaches	Problems of ownership and right of usage
Increase of job opportunities to local community to work in several remediation processes	Increase of industrial pollution
Improvements in legality and policy which was started with the EU adaptation processes	Increase of using chemical pesticides and fertilizers
Developments in the watershed management in the world.	Threats on the biological diversity
Increase of scientific research capacities.	Inadequacies in the terms of the legal sanction about problematic application processes Negative effects on watershed which is caused by climate change process Increase of industry at the watershed areas Increase of mining at the watershed areas

Figure 3. SWOT Analysis for Watershed Management in Turkey (Anonymous, 2013).

DISCUSSIONS AND CONCLUSION

All the organizations is interested with poor people and dedicated to eradicating rural poverty in developing countries.

Main problem about this issue is rapid demographic chance in both rural and urban areas in developing countries. Rural areas have been emptied by the migration progress that is caused by unemployment and poverty in rural areas. Beside this situation, huge urban sprawl has been seen on many cities in developing countries. Water demand on the cities that has got huge reservation of the population has been become big problem for clean water resources and sustainability of this resources can not be produced by the managers and governments. However, watershed management should be collaborative. Collaborative watershed partnerships consist with different land use patterns: one in an urban environment, and the other rural. A good practice of healthy and sustainable watershed management is based on a good understanding of these two distinct territories that have different organic and fabric organization schema.

Researchers have paid particular attention to the effect of land use on water quality (Lenat and Crawford, 1994; Hall *et al.*, 1994, Wang and Yin, 1997), a water-quality component often is missing in land-use plans and land-use planning is rarely used in water-quality management. This could be because water-quality management and land use planning often are administrated by different agencies that do not coordinate constantly. Both of planning agencies and local authorities are not use feedback datas for developing plans to monitor of extensive land use and water-quality chances. Water-quality management agencies traditionally address existing water-quality problems rather than to prevent to cause a problem in the main sources. Planners and decision-makers should pay proper attention to water-quality issues in evaluating plans and facilitating collaborations. They should pay enough attention to significancy of land and water collaboration (Wang and Yin, 1997),.

Governments and development organizations worldwide are searching for new ideas on how to bring more participation that is public into environmental policy decision processes. Some social and political movements, too, are expanding their participatory outreach and organizing techniques (CIELAP, 2004). Many of the nongovernmental organizations and green activities seek to focus public opinion globally on the importance of water rights (Council of Canadians, 2006). In addition, as mentioned above, the European Water Framework Directive is pushing jurisdictions and cross-jurisdictional water basin committees to implement new participatory processes. Researchers, consultants and activists are generating practices, which can be widely discussed and shared. However, in the end, while many insights can come from hearing what worked and what did not in other places/situations, there is no substitute for locally designed and locally appropriate public participation processes, both within and outside of government (David, 2005).

As the conclusion of the European Water Framework Directive Guidance Document on Public Participation states, "The preamble of the Water Framework Directive includes a very clear statement: active public involvement is most likely the key to success with regard to achieving the desired water quality objectives." In other view, the water users and water polluters need to be turned into part of the solution, not kept outside the considerations as part of the problem. Active involvement is important, however, to take into account that no blueprint solution can be provided. Each River Basin District has to find its own way to handle this, taking into account the prevailing cultural, socio-economic, democratic and administrative traditions. Careful planning, stakeholder analysis is a particular recommendation, but each competent authority has to accept that a dynamic and learning process based on "trial and error" is the challenge to embark on. Experiences showed that given sufficient time it would pay off in the long run (EC,). According to Perkins (2011), government initiatives in a very positive light and ignores issues such as in what sense water users who are unserved by infrastructure or too poor to pay for water are really "stakeholders," how the "payoff" of a public learning process might be measured, and to whom this payoff accrues (Perkins, 2011).

Watershed Watershed management project results which are successful in the term of both theory and application, the role of class and gender, among other differences, as determinants of everyone's standpoint and possibilities for participation must be acknowledged. Liberal individualism is certainly not the only thing going on in any participatory process. The truly radical nature of participation only appears as and when it leads to economic redistribution – not just policy-making within existing structures of distribution. Under capitalism and due particularly to its second, ecological, contradiction (O'Connor, 1994), the pressing need for local environmental knowledge and the contributions of diverse constituencies (Fischer, 1993; McKinney, 2002) in order to address ecological constraints has arguably loosened the controls which the state has traditionally placed on democracy. The type of policy analysis that is "antagonistic to authentic democratic participation" (Fischer, 1993) is now giving way in some instances to more democratic policy-generating processes as a means of addressing intractable environmental problems. However, the inherent open questions of gender, "race," ethnicity and class inequities remain. To address both the "top-down" and the "bottom-up" challenges to broadening public involvement in watershed decision processes as cited above, a creative combination of grassroots environmental education and community organizing is needed. Community-based environmental education initiatives, which are relevant and they are interesting for local residents and increase their knowledge of watershed issues (Fischer, 1993).

Watershed affect the daily lives of every one of many countries 's citizens and provide a powerful wall of protection for countries 's economic development. Data are scarce, but subjective understanding is clear: throughout much of the developing world watersheds are vital to the livelihoods of many millions of people.

Today issues about the water management should be appropriate to the nature as Thales had been said "water is not a just a common material besides it is a spiritual basin". Because of them water management should not be just an engineering work besides it should be a social work which should consider on the both social and cultural work too. So that, common problems can be solved easily and outputs of the projects will be more productive. Rivers are not just a natural sources for transportation, fresh water, energy or irrigation, none or the less they are the basic ground for human culture of the nearby areas. All the ancient great civilizations have been settled near the great rivers, and most of the modern civilization have a peace with their rivers. Because of them combining links between nature and society is the most important issue for sustainability.

REFERENCES

1. Anonymous, 2009. White Paper Holistic Water Resources Management: A Professional Education Plan, H2C1 Team Mississippi State Universityhttp://www.gri.msstate.edu/ publica-tions / docs /2009 /09 /6310 WatershedCourseWP2 Sep09.pdf, Date Accessed: 21.07.2012
2. Anonymous, . http://en.wikipedia.org/wiki/Thales Date Accessed: 21.07.2012
3. Anonymous, . http://en.wikipedia.org/wiki/Heraclitus Date Accessed: 21.07.2012
4. Anonymous, . http://www.worldbank.org/projects/009023 eastern-anatolia-watershed-project?lang=en, Date Accessed: 21.07.2012
5. Anonymous, . http://damocracy.org/?page_id=417. Date Accessed: 21.07.2012
6. Anonymous, . http://damocracy.org/wp- content/ uploads/2012/04/hasankeyf_hasankeyfesadakat2.jpg, Date Accessed: 21.12.2012
7. Anonymous, . http://en.wikipedia.org/wiki/File:Hasankeyf.JPG, Date Accessed: 28.12.2012
8. Anonymous, . http://www.ifad.org/governance/index.htm, Date Accessed: 11.08.2012
9. Anonymous, . http://www.fao.org/about/en/ Date Accessed: 11.08.2012
10. Anonymous, . http://www.iwmi.cgiar.org/About_IWMI/ Overview.aspx Date Accessed: 11.08.2012
11. Anonymous, 2013. Ministry of Forestry and Hydraulic Works, 2012. Strategy of National Watershed Management Report. Date Accesses: 10.04.2013. http://www.ormansu.gov.tr/osb/haberduyuru/duyurular/12-04-19/Ulusal_Havza_Y%C3%B6netim_Stratejisi_Tasla%C4%9F%C4%B1.aspx?sflang=tr
12. B. Benham, K. Brannan, T. Dillaha, S. Mostaghimi, G. Yagow, 2002TMDLs (total maximum daily loads)-terms and de?nitions, Blacksburg, VA: Virginia Tech.

13. J. R. Blatt, 1993The anti-environmental backlash: The wise use movement's in?uence on river protection in New England. Unpublished master's thesis. Tufts University, Medford, Massachusetts.

14. D Bossio, ., K Geheb, ., 2008. Conserving land, protecting water. Comprehensive Assessment of Water Management in Agriculture Series 6. Wallingford, UK: CAB International; Colombo, Sri Lanka: International Water Management Institute (IWMI); Colombo, Sri Lanka: CGIAR Challenge Program on Water and Food.http://www.iwmi.cgiar.org/Publications/CABI_Publications/CA_CABI_Series/Conserving_Land_Protecting_Water/protected/9781845933876.pdf

15. B. Bowden, 1999Integrated catchment management rediscovered: An essential tool for a new millennium. In: Proceedings of Manaaki Whenua Conference: A Three Day Conference on Science for Resource Management. 2123

16. J. W. Bulkley, 1995Integrated watershed management: past, present, and future. Water resources management, 100, 718

17. R. Chambers, A. Pacey, L. A. Thrupp, 1989Farmer first: Farmer innovation and agricultural research.London: Intermediate Technology Publications.

18. C. Chess, G. Gibson, 2001Watersheds are not equal: exploring the feasibility of watershed managementJ Am Water Resour Assoc 2001;37477582

19. CIELAP2004Public Participation in Great Lakes Management. Canadian Institute for Environmental Law and Policy, Toronto.

20. B. A. Cigler, 1998Emerging trends in state-local relations. In R. L. Hanson (Ed.), Governing partners: State-local relations in the United States. Boulder, CO: Westview Press.

21. J. Cobourn, 1999Integrated watershed management on the Truckee River in NevadaJournal of the American Water Resources Association 353623632

22. P. V. Combe, K. Najjar, 2009Framework for Sustainable Watershed Management in Pocono Creek Watershed, Delaware River Basin Commission, West Trenton, New Jersey

23. S. M. Cormier, M. Smith, S. Norton, 2000Assessing ecological risk in watersheds: a case study of problem formulation in the

Big Darby Creek watershed, Ohio, USA. Environ Toxicol Chem 19210821096

24. Council of Canadians, 2006. Blue Planet Project. <http://www.blueplanetproject. net/Movement/index.html>. http://www.canadians.org/

25. S. Darghouth, C. Ward, G. Gambarelli, C. Styger, J. Roux, 2008Watershed Management Approaches, Policies, and Operations: Lessons for Scaling up watersector board siscussion papaer series, Paper 11The World Bank, Washington, DC.

26. E.T David, ., 2005. Developing Watershed Management Organizations in Pilot Sub-Basins of the Ping River Basin, Project Report of Participatory Watershed Management for the Ping River Basin, Thailand, url: http://www.worldagroforestry.org/sea/Publications/files/report/RP0129063 .pdf, accessed date: 01.02.2013.

27. M. C Deauna, ., F Lamayo, ., 2000. Introduction to analytic geometry and calculus- Quezon City: SIBS Pub. House, 9-71791-073-1//cristina327.hubpages.com/hub/History-of-Geometry

28. M. C. Dorf, C. F. Sabel, 1998A constitution of democratic experimentalismColumbia Law Review982267473

29. EC, 2002. Guidance on Public Participation in Relation to the Water Framework Directive: Active Involvement, Consultation, and Public Access to Information. EC Circa Website. <http://forum.europa.eu.int/Public/irc/env/wfd/library?l=/framework_directive/guidance_documents/participation_guidance&vm=detailed&sb=Title >(accessed10.07.04).

30. S. Ewing, 1999Landcare and community-led watershed management in Victoria, AustraliaJournal of the American Water Resource Association 353663674

31. F. Fischer, 1993Citizen participation and the democratization of policy expertise: from theoretical inquiry to practical casesPolicy Sciences165187

32. R. E. Freeman, 1984Strategic management: A stakeholder approach.Boston: Pitman.

33. D.P Garrity, ., F.K Akinnifesi, ., O.C Ajayi, ., S.G Weldesemayat, ., J.G Mowo, ., A Kalinganire, ., M Larwanou, ., J Bayala, ., 2010. Evergreen agriculture: a robust approach to food security in

Africa. Garrity D.P., Akinnifesi F.K., Ajayi O.C., Weldesemayat S.G., Mowo, J.G., Kalinganire A., Larwanou M., Bayala J., 2010. Evergreen agriculture: a robust approach to food security in Africa. Food Security . http://dx.doi.org/10.1007/s12571-010-0070-7 2 3 197214 . http://dx.doi.org/10.1007/s12571-010-0070-7

34. S. Giri, P. Nejadhashemi, S. A. Woznicki, 2012Evaluation of targeting methods for implementation of best management practices in the Saginaw River Watershed

35. J. Glicken, 2000Getting stakeholder participation 'right': a discussion of the participatory processes and possible pitfalls. Environmental Science and Policy, 3305310

36. B. F. Golden, 1998Issues in developing and implementing a successful multiparty watershed management strategy. 353368in R. J. Reimold (ed.), Watershed management: Practice, policies and coordination. McGraw-Hill, New York.

37. N. S. Grigg, 1998Coordination: the key to integrated water management. Water resources. Update, Special Issue 111Carbondale, IL:Universities Council on Water Resources.

38. A. Groetschel, I. Müller-neuhof, I. Rathmann, H. Rupp, X. Santillana, A. Söger, J. Werner, 2000Watershed Development in Gujarat- A problem-oriented survey for theIndo-German Watershed Development Programme- Humboldt- Universität Zu Berlin Landwirtschaftlich-gärtnerische Fakultät Sle Centre for Advanced Training in Rural Development Ahmedabad/Berlin.

39. J. Gutrich, D. Donovan, M. Finucane, W. Focht, F. Hitzhusen, S. Manopimoke, D. Mccauley, B. Norton, P. Sabatier, J. Salzman, V. Sasmitawidjaja, 2005Science in the public process of ecosystem management:lessons from Hawai, Southeast Asia, Africa and the US Mainland. Journal of Environmental Management, Aug:(763197209

40. R Hajjar, .; D.I Jarvis, .; B Gemmill-Herren, ., 2008. The utility of crop genetic diversity in maintaining ecosystem services. Agriculture Ecosystems & Environment 123 261270 . http://dx.doi.org/10.1016/j.agee.2007.08.003

41. L. W. Hall, S. A. Fischer, W. D. Killen, Jr,Scott M. C., Ziegenfuss M. C., Anderson R. D., 1994Status assessment in acid-sensitive

and non-acid-sensitive Maryland coastal plain streams using an integrated biological, chemical, physical, and land-use approach. Journal of Aquatic Ecosystem Health 3145167

42. S. D. Hardy, T. M. Koontz, 2010Collaborative watershed partnerships in urban and rural areas: Different pathways to success? Landscape and Urban Planning 957990

43. HeathcoteIsobel W., 1998Integrated Watershed Management: Principles and PracticesJohn Wiley& Sons, Inc. New York.

44. C. L. Lant, 1999Introduction human dimensions of watershed managementJournal of the American Water Resources Association483486

45. D. R. Lenat, and J. K. Crawford, 1994Effects of land use on water quality and aquatic biota of three North Carolina Piedmont streamsHydrobiologia294185199

46. B.M Liu, .; Y Abebe, .; O.V McHugh, .; A.S Collick, .; B Gebrekidan, .; T.S Steenhuis, . 2008. Overcoming limited. information through participatory watershed management: Case study in Amhara, Ethiopia. Physics and Chemistry of the Earth 33 (1-2): 13-21.http://dx.doi.org/10.1016/j.pce.2007.04.017

47. M. Mccartney, V. Smakhtin, 2010Water storage in an era of climate change: addressing the challenge of increasing rainfall variability. IWMI BluePaper. Colombo: International Water Management Institute.

48. McKinneyMatthew, Harmon, W.,2002Public participation in environmental decision making: is it working? National Civic Review 91 (2), 149.

49. S Michael, . 2001. Making Collaborative Watershed Management Work: The Con?uence of State and Regional Initiatives. Environmental Management 27 1 2735 DOI: 10.1007/s002670010131

50. B. Mitchell, Integrated water management: international experiences and perspectives.London, Belhaven Press. Mitchell, R.K., Agle, B.R. & Wood, D.J. (1997Toward a theory of stakeholder identification and salience: Defining the principle of who and what really counts. Academy of management review, 22 (4), 853886

51. R. K. Mitchell, B. R. Agle, D. J. Wood, 1997Toward a theory of stakeholder identification and salience: Defining the principle of who and what really counts.Academy of management review, 22 (4), 853886

52. D. Molden, R. Tharme, I. Abdullaev, R. Puskur, 2007Irrigation. In Scherr, S.J.; McNeely, J.A. eds. Farming with nature:The science and practice of ecoagriculture. Washington, DC, USA: Island Press. 231249

53. National Research Council1999New Strategies for America's WatershedsNational Academy Press, Washington, DC.

54. O. Connor, James, 1994Is sustainable capitalism possible? In: O'Connor, Martin (Ed.), Is Capitalism Sustainable? Guilford Press, New York, 152175

55. L. H Padma, . M Lim-Applegate, . Scoccimarro. 2001. The adaptive decision-making process as a tool for integrated natural resources management: Focus, attitudes, and approach. Conservation Ecology 5(2):11. http://www.consecol.org/5 iss2/ art11

56. H. C. Pereira, 1973Land use and water resources in temperate and tropical climatesLondon, Cambridge University Press.

57. E. Perkins, 2011Public participation in watershed management: International practices for inclusiveness/Physics and Chemistry of the Earth 362011204212

58. R. E. Rhoades, 1999Participatory Watershed Research and Management: Where the Shadow Falls.Gatekeeper Series 81London: IIED.

59. G. Said, D. K. Sehlke, T. Stevens, D. Glover, W. Sorensen, T. Walker, Hardy, 2006Exploring an innovative watershed management approach: From feasibility to sustainability,Energy 31200623732386

60. D. H. Scott, T. M. Koontz, 2010Collaborative watershed partnerships in urban and rural areas: Different pathways to success? Landscape and Urban Planning 9520107990

61. A. N. Sharpley, 2000Agriculture and phosphorus management: The Chesapeake Bay and Boca Raton FL. Lewis Publishers, London.

62. A. K. Singh, T. J. Eldho, and D. Prinz, 2002Integrated watershed management approach for combating drought in a semi-arid region in India: The case of Jhabua watershed. In 2nd World Water Congress: Integrated Water Resources Management 8592Alliance House, London.

63. StahlRG; Bachman, RA; Barton, AL; et al. (1999A multi-stakeholder framework for ecological risk management: Summary of a SETAC technical workshop. Environ Toxicol and Chem 182122

64. J. E. Steiguer, J. Duberstein, V. Lopes, 2003The Analytic Hierarchy Process as a Means for Integrated Watershed Management, Proceedings of the 1st Interagency Conference on Research on the Watersheds, Oct. 27-30, Agricultural Research Service, Benson, Arizona,736740

65. B. Swallow, R. Meinzen-dick, 2009Payment for Environmental Services: interactions with property rights and collective actionIn: Beckmann V.; Padmanabhan, M. eds. Institutions and Sustainability: Political Economy of Agriculture and the Environment. Springer. 243265

66. M.J Swift, .; M.N Izac, .; M van Noordwijk, . 2004. Biodiversity and ecosystem services in agricultural landscapes- are we asking the right questions? Agriculture, Ecosystems & Environment 104 1 113134 . http://dx.doi.org/10.1016/j.agee.2004.01.013

67. M. Tiffen, M. Mortimore, and F. Gichuki, 1994More People, Less Erosion, Environmental Recovery in Kenya,African Centre for Technology Studies, Nairobi, Kenya, 326.

68. J Tournebizea, ., B Vincenta, , C Chaumonta, , C Gramagliab, , C Margoumc, , P Mollec, , N Carluerc, , J. J Grilc, , 2011. Ecological services of artificial wetland for pesticide mitigation. Socio-technical adaptation for watershed management through TRUSTEA project feedback, Procedia Environmental Sciences 9 2011 183190 doi:10.1016/j.proenv.2011.11.028

69. U. S. Epa, 1991Guidance for water quality-based decisions: the TMDL process, EPA 444/4-91-001. Washington, DC.

70. U.S EPA, , 1992 A framework for ecological risk assessment. Risk Assessment Forum, Washington, DC; EPA/630/R-92/001. Available online athttp://www.epa.gov/ncea/raf.

71. U.S EPA, , 2001. Improved science-based environmental stakeholder process, a commentary by the EPA science advisory board. EPA Science Advisory Board, Washington, DC; EPASAB-EC-COM-01006 . Available online at http://epa.gov/sab/pdf/eccm01006.pdf.

72. U.S EPA, , 2012. What is a watershed? http://water.epa.gov/type/watersheds/whatis.cfm

73. University Outreach& Extension Water Quality Program (UOEWQP), 2004. Fostering Locally-Led Holistic Watershed Managementhttp://www.usawaterquality.org/conferences/2004/posters/TharpMO.pdf

74. A. J. Van Der Zijpp, J. A. J. Verreth, Le Quang Tri; van Mensvoort, M.E.F.; Bosma, R.H.; Beveridge, M.C.M. 2007Fishponds in farming systemsWageningen Publishers, Netherlands.

75. World Bank, 2003. China Loess Plateau Watershed Rehabilitation Project Report. Report 25701 http://www-wds.worldbank.org

76. I. W. Wamalwa, 2009Prospects and limitations of integrated watershed management in Kenya: a case study of mara watershed, Lund university international masters programme in environmental studies and sustainability science, A Master Thesis, Sweden 51

77. X. Wang, and Z. Yin, Y. 1997Using GIS to assess the relationship between land use on water quality at a watershed level. Environment International23103114

78. X. Wang, 2001Integrating water-quality management and land-use planning in a watershed contextJ. Environ. Manage. 61 (1), 25 EOF36 EOF

79. R.J Zomer, .; A Trabucco, .; R Coe, .; F Place, . 2009. Trees on farm: analysis of global extent and geographical patterns of agroforestry. ICRAF Working Paper 89 Nairobi: World Agroforestry Centre. http:// www. worldagroforestrycentre.org/downloads/ publications/PDFs/WP16263.PDF

Chapter 4

Functional Changes in Littoral Macroinvertebrate Communities in Response to Watershed-Level Anthropogenic Stress

Katya E. Kovalenko[1,2]*, Valerie J. Brady[2] , Jan J. H. Ciborowski[1] , Sergey Ilyushkin[3] , Lucinda B. Johnson[2]

[1]University of Windsor, Windsor, Ontario, Canada,

[2]Natural Resources Research Institute, University of Minnesota Duluth, Duluth, Minnesota, United States of America,

[3]Colorado School of Mines, Golden, Colorado, United States of America

ABSTRACT

Watershed-scale anthropogenic stressors have profound effects on aquatic communities. Although several functional traits of stream macroinvertebrates change predictably in response to land development and urbanization, little is known about macroinvertebrate functional responses in lakes. We assessed functional community structure, functional diversity (Rao's quadratic entropy) and voltinism in macroinvertebrate communities sampled across the full gradient of anthropogenic stress in Laurentian Great Lakes coastal wetlands. Functional diversity and voltinism significantly decreased with increasing development, whereas agriculture had smaller or non-significant effects. Functional community structure was affected by watershed-scale development, as demonstrated by an ordination analysis followed by regression. Because functional community structure affects energy flow and ecosystem function, and functional diversity is known to have important implications for ecosystem resilience to further environmental change, these results highlight the necessity of finding ways to remediate or at least ameliorate these effects.

INTRODUCTION

Widespread anthropogenic modification of the landscape is jeopardizing freshwater ecosystems and their services. Watershed-level anthropogenic stress has negative effects on macroinvertebrate communities in lakes, streams and wetlands [1],[2],[3]. Development in the watershed is characterized by reduction/removal of natural vegetation, increased road density and increased proportion of other impervious surfaces, as well as higher human population density. The effects of development-associated stressors and those associated with agricultural land use on macroinvertebrate communities are mediated through increased nutrient loading, point and non-point source pollution, sediment loads, altered hydrologic and temperature regimes, and habitat destruction and fragmentation in riparian zones and littoral areas [1].

Several functional characteristics (*e.g.*, flow, drag or silt adaptations, respiration and locomotion techniques, feeding habits, voltinism, *reviewed in* [3]) are considered to be important indicators of the state of an ecosystem and its potential resilience to further anthropogenic modification. In particular, functional diversity (FD) is a critical property of a group of organisms at any scale because increased trait space breadth is likely to be associated with a greater diversity of ecosystem processes and nutrient pathways, which in turn increase resistance and resilience to perturbations [4],[5]. FD can be related to reticulation of the food web, which has important implications for the resilience of food webs as demonstrated in a theoretical study [6]. In terrestrial plant assemblages, greater functional diversity has been shown to maximize resource use in heterogeneous environments, and affect energy flow and ecosystem function (*reviewed in* [4], [7], [8]), and similar patterns were observed in theoretical studies [9]. However, little is known about the effects of reduced FD in littoral systems. Another important functional trait, voltinism, may have important implications for temporal redistribution of nutrient processing and ecosystem stability, and proportion of univoltine and other longer-lived organisms was shown to decrease with increasing land use stress [10]. Furthermore, changes in the relative abundance of different invertebrate functional groups can alter nutrient processing characteristics [11], [12], potentially triggering cascading effects in higher trophic levels, which may also affect littoral-pelagic and aquatic-terrestrial habitat coupling, by virtue of this group's central position in aquatic food webs.

Although invertebrate functional trait responses to increasing anthropogenic stress have been described in streams-particularly changes in voltinism and respiration type [3] - those findings may not apply to lake and wetland systems because the effects of watershed-scale stressors are no longer mediated primarily through hydrological alteration and canopy clearing [1]. In addition, some of the more sensitive functional attributes, such as the proportion of semi- and merovoltine taxa and anoxia-intolerant taxa, are already under-represented in lentic systems because these systems routinely experience greater variability in parameters such as dissolved oxygen, temperature, and other factors [13]–[15]. It is important to understand whether functional attributes of littoral macroinvertebrate communities change in response to watershed-scale anthropogenic stress in order to design management and conservation strategies specific to freshwater littoral systems.

To assess the functional responses of lentic macroinvertebrates, we used wetland macroinvertebrate community composition data from the Great Lakes Environmental Indicators project [16] collected across a full gradient of watershed development stress (basin-wide minimum to maximum) to 1) test whether FD and the relative abundance of longer-lived (uni-, semi- and merovoltine) organisms are affected by watershed-level development and agriculture, and 2) investigate the functional changes in invertebrate communities. The *a priori*prediction was that FD and proportion of longer-lived taxa would decline with increasing development and agriculture due to an overall decrease in taxonomic diversity and reduced habitat availability resulting from the combined effects of direct habitat degradation and pollution commonly associated with these anthropogenic stressors [1]. Due to the large number of taxa and their wide range of physiologic tolerances and habitat requirements, macroinvertebrates can serve as an indicator of the changes impacting other assemblages, as well as a warning signal of changes in littoral food webs.

MATERIALS AND METHODS

Macroinvertebrates were sampled in 101 coastal wetlands of the U.S. coastline of the Laurentian Great Lakes in the summer of 2002 and 2003 (Fig. 1). This dataset was previously collected for a different purpose, as part of a multidisciplinary effort to identify indicators of anthropogenic stress in the Great Lakes coastal zone (the Great Lakes Environmental Indicators project). No permits were required for macroinvertebrate sampling by the U.S. at the time of the original study and no recognized endangered or threatened invertebrate taxa occur in those coastal wetlands. All appropriate protected area sampling permits were secured by the original study.

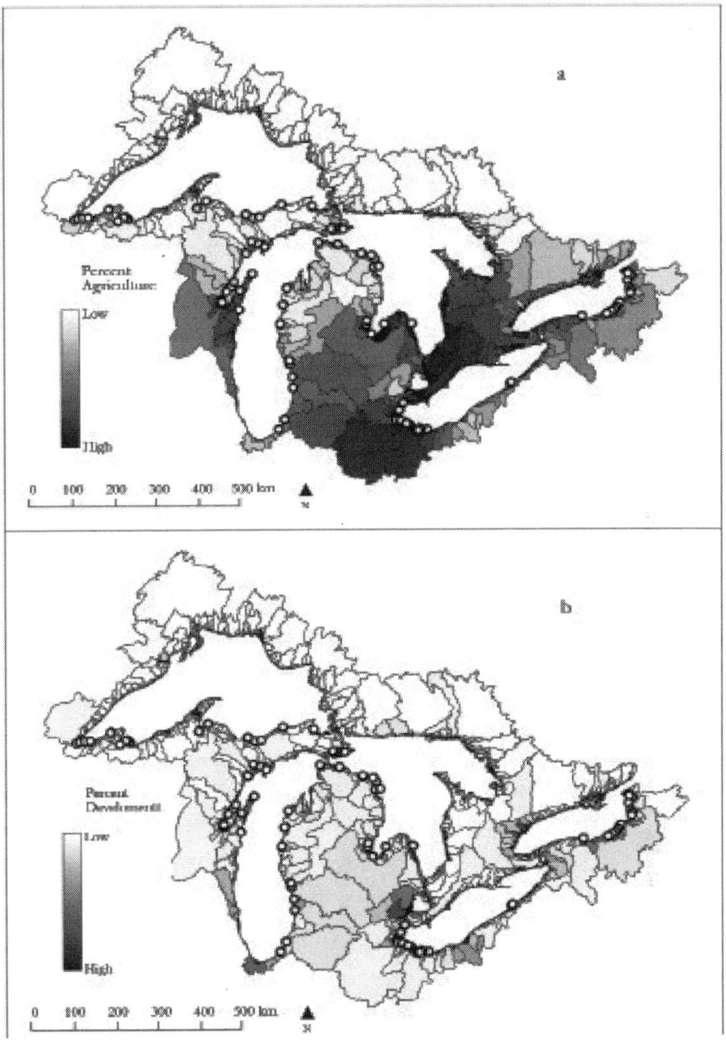

Figure 1. A map of the study sites in the coastal wetlands of the Laurentian Great Lakes, overlaid with the land-use stressor gradient for a) agriculture and b) development.

The Laurentian Great Lakes include five glacial till lakes: Superior, Michigan, Huron, Erie and Ontario, located in the temperate part of North America. Lakes range from oligotrophic to eutrophic, with the surrounding land use spanning a gradient from completely unimpacted to highly impacted by anthropogenic activities including development and agriculture. Wetlands were equally distributed among lakes and across a full gradient of anthropogenic stress, previously defined as a composite of five major classes of anthropogenic pressure: agriculture, atmospheric deposition, land cover, human development, and point source pollution. The site selection procedure ensured representation of four geomorphic wetland types, including riverine, barrier-beach protected, and lacustrine coastal wetlands and embayments [17]. All wetlands were hydrologically connected to a Great Lake, and most wetlands had well-developed submerged and emergent macrophyte communities.

Proportion of human development and agriculture (by area) in a wetland's watershed was derived from the USGS National Land Cover Dataset (2001) for each catchment delineated using ArcHydro with 10-m Digital Elevation Models (see [18] for details). Development was defined to include all residential, commercial and industrial areas, but did not include road density.

Invertebrate sampling and functional metrics

To ensure the most complete sampling of the different habitats present within a wetland, macroinvertebrates were collected from all representative near shore land-use and shoreline (e.g., sand beach, cobble beach, lawn, etc.) zones in each wetland. In each land-use-shoreline zone, two transects were extended perpendicular to shore, and macroinvertebrates were collected using 30-second D-frame dipnet sweeps at 0.25 and 0.75-m water depths along each transect. Sweeps were done through the water column from the bottom to the surface, in a forward direction parallel to the shore, regardless of vegetation type or presence. Samples were rinsed in a 250 μm sieve net or bucket to remove fine particles, and preserved in Kahle's solution for laboratory processing. In the laboratory, macroinvertebrates were identified to the highest possible resolution (genus for most insects) using the most current keys available at that time [19], [20]. Data from the two transects in each zone were first averaged by depth, then averaged across zones to achieve site-level data. After taxonomic identification and sample averaging were completed, traits were assigned to each taxon using the latest reviews[14], [19], [21]–[27] and expert judgment (see Table S1 for details).

FD was measured using Rao's Quadratic entropy (Q), which accounts for the relative abundances of species and for the functional differences between species by measuring differences between two randomly selected individuals with replacement (*see* [28], [29] for formulas and performance evaluation). It is closely correlated with the index of functional dispersion based on the centroid-distance approach [30], although it has also been demonstrated that this metric can be conservative under some scenarios due to negative covariation with species richness [29]. This analysis was performed with trait variables including trophic status, feeding mechanism, locomotion and primary and secondary functional feeding groups (Table S2), because those were the traits for which information was most consistently available across all encountered taxa. All traits were combined into a single Q-space.

The voltinism measure was expressed as the proportion of a sample comprised of taxa with long-lived aquatic phases (i.e. the proportion of individuals in a sample that were uni-, semi or merovoltine) to all other taxa. Voltinism is defined only for insect taxa; this life-history information was available for 77 taxa and unavailable for the remaining 85 insect taxa (of the total of 222 insect and non-insect taxa). This was mostly due to incomplete knowledge, as voltinism is one of the most difficult traits to describe, and to a lesser extent due to the level of resolution (variable life history at the species level, but identification to genus level) and differences in life history across a large geographic range of a taxon. Although all studies looking at voltinism unavoidably have this limitation, we chose to analyze this trait because it is a very important indicator of the state of macroinvertebrate assemblages and there is no *a priori* reason to suspect that the effect of this missing information is directional and not conservative on our ability to understand how longer-lived insects are affected by land use. Fifty-eight taxa in our samples were uni-, semi- or merovoltine, primarily belonging to Odonata and several genera of Trichoptera and Ephemeroptera, along with a few rarely occurring groups.

Statistical Analyses

Multiple regression (MR) analyses were used to test for relationships between watershed stressor variables and voltinism and FD. Percent development was log-transformed, whereas percent agriculture was not transformed to satisfy the assumptions of regression residuals distribution, as tested using Shapiro-Wilk tests. To address the possibility of the confounding effect of sampling across large geographic scales, we conducted additional multiple regression analyses using latitude as a third predictor for each of the response variables. Latitude was significantly, but weakly correlated with agriculture (linear regression permutation $p = 0.010$, $R^2 = 0.06$), but not with development ($p = 0.26$). Functional community structure was summarized using Principal Components Analysis (PCA) conducted on the log-transformed relative abundances of 12 functional traits. The resulting site Principal Component scores for each factor were regressed against the predictor variables as described above. Analyses were done in R 2.12.2 (R Development Core Team, Vienna, Austria), using packages *vegan* and *lattice*. Package *relaimpo* was used to calculate the "*lmg* relative importance metric" [31] for predictors in the multiple regression models, which produces averages of sequential sums of squares over all orderings of regressors (*see* [31] and package documentation for details). Although computationally intensive, the *lmg* procedure has been recommended because it decomposes R^2 into non-negative contributions and accounts for direct effects as well as adjustments for other regressors in the model [32]. Natgrid, a two-dimensional random data interpolation package, was used to resample data to a regular grid and produce contour plots. Natgrid, based on nngridr package [33], implements a natural neighbour interpolation method and uses a weighted average method. The package was implemented through *matplotlib* [34], plotting library for the Python programming language.

RESULTS

Watershed development exerted stronger effects on all functional response variables than did watershed agriculture (Fig. 2). Rao's functional diversity was significantly reduced by both development and agriculture (Fig. 3, MR $F_{2, 97} = 9.60$, adjusted $R^2 = 0.16$, $p = 0.003$ and 0.004 for development and agriculture, respectively; see Table S3 for details and all regression coefficients). Development was the more important driver of this relationship (Fig. 2, Fig. S1a). To illustrate the extent of this effect, sites with more than 10% development had a

25% reduction in functional diversity compared to the less developed sites. The latitudinal predictor did not contribute significantly to the MR (p = 0.58), and its presence had almost no effect on the relative importance of the two stressors, confirming the lack of confounding effects (Table S3).

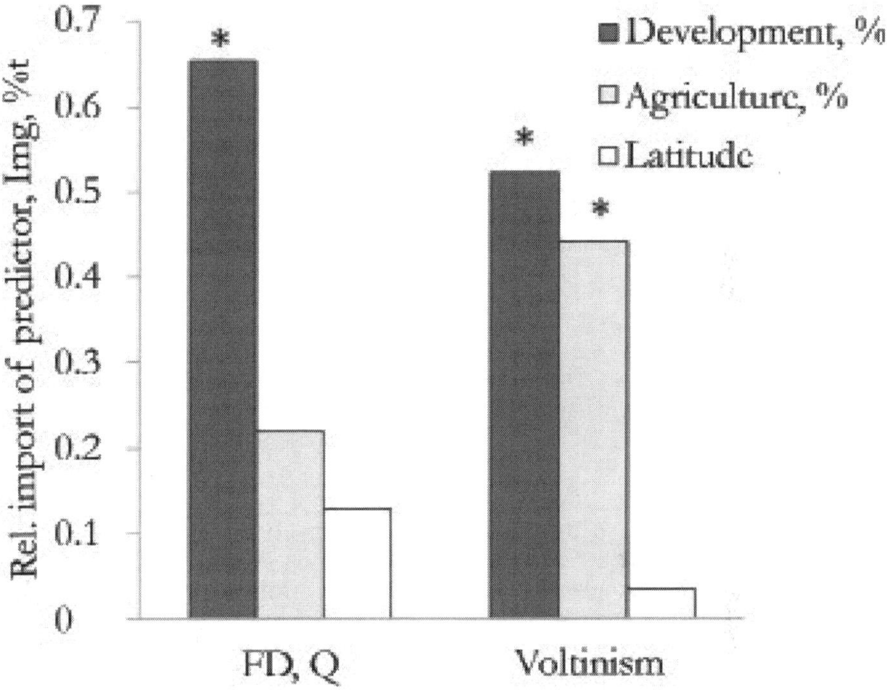

Figure 2. Relative importance of the two watershed stressors and latitude in explaining variation in the functional attributes of macroinvertebrate communities, as proportional *lmg* contribution. Note the overriding importance of development over agricultural stress and the lack of significant geographic confounding. Asterisks indicate predictors that were significant in the multiple regression (p<0.05).

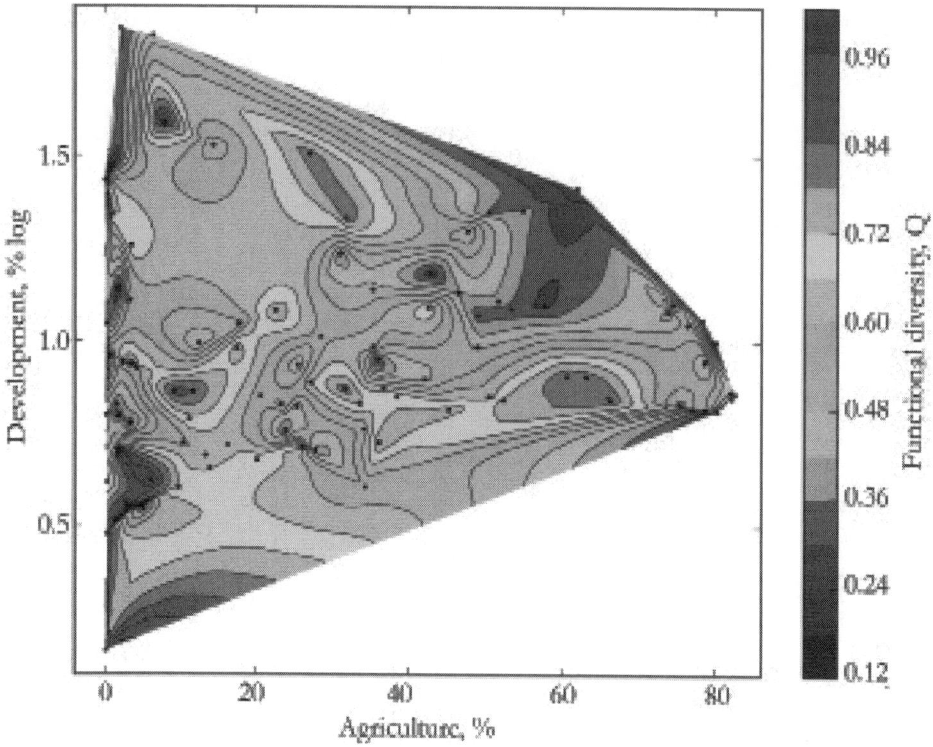

Figure 3. Macroinvertebrate functional diversity (Rao's Q) as a function of development and agricultural stress in the watershed. Note that maximum values of the functional trait are observed mostly below a certain proportion of development (around mid-point of the y-axis, corresponding to 10% un-transformed % development), indicating the over-riding contribution of that stressor; and if high values are observed at the higher levels of development, those occur only at sites with minimal agriculture. *[figure footnote] Contour lines divide the figure into a region where values are higher than they are on the contour line itself and a region where they are lower; values change across the line but not along the contour line, and the gradient is larger where contour lines are packed closer together.

Relative abundance of those longer-lived taxa declined with increasing development in the watershed (MR $F_{2, 97} = 11.17$, adjusted $R^2 = 0.17$; Fig. 4, $p<0.001$ for development, also Fig. S1b). The weak (Fig. 2) but significant negative effect of agriculture observed in the two-factor model ($p = 0.017$) was no longer significant ($p = 0.062$) when latitude was included as a predictor (see Table S3 for model details and regression coefficients).

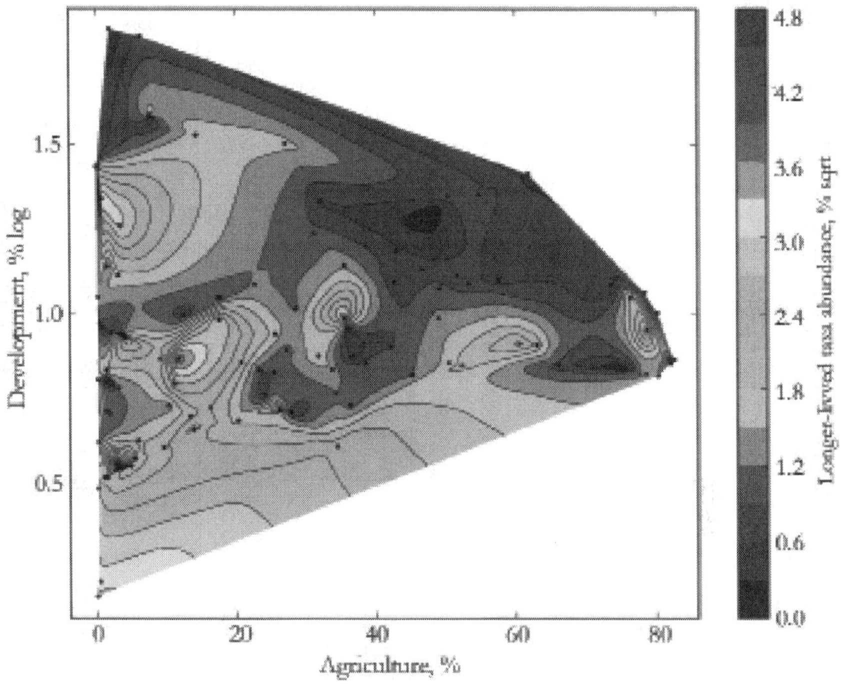

Figure 4. Relative abundance of long-lived taxa (uni-, semi- and merovoltine) as a function of development and agricultural stress in the watershed.

The overall functional community structure was weakly but significantly affected by the extent of development in the watershed. The first PC axis, explaining 32% of the variation in functional community structure (eigenvalue = 3.74), was driven primarily by burrowers and filterers on one end vs. scrapers and clingers on the other end (*see* Table S4 for loadings). This axis was significantly positively correlated with percent development (MR $F_{2, 97} = 5.82$, adjusted $R^2 = 0.11$; $p = 0.004$) but not with agriculture. In other words, the proportions of positively-loading groups, including burrowers and filterers, were positively correlated with greater amounts of watershed development, whereas scrapers and clingers were negatively correlated. This axis score tended to be negatively correlated with agriculture, although not significantly ($p = 0.053$), with this trend disappearing when we added latitude as a predictor, similar to the pattern observed for other response variables. The second PCA axis (eigenvalue = 3.10, percent variation explained = 19%) was mostly driven by omnivores (Table S4). This axis was weakly negatively correlated with percent agriculture ($p = 0.011$, $R^2 = 0.06$).

DISCUSSION

This study demonstrates that macroinvertebrate functional diversity and abundance of longer-lived taxa in Laurentian Great Lakes coastal wetlands decreased significantly with greater watershed-level stress. Although macroinvertebrate communities are influenced by factors acting at many spatial scales, as has been demonstrated in a number of stream studies (e.g.[3], [35], [36] but see [37] for lake margins), our results indicate that the negative effects of watershed-level development were sufficiently robust to be detected against the background of a strong geographic gradient, the presence of additional stressors and a wide range of local habitat features. Compin and Céréghino [38] used self-organizing maps to show that, in human-modified landscapes, effects of stream watershed-scale stressors over-rode the influence of natural physical factors. It also means that large-scale stressors can produce detectable changes in ecosystem function, at least with respect to macroinvertebrate functional traits. The effect of watershed development was significant, but not highly predictive, reflecting the hierarchical complexity of the underlying factors affecting macroinvertebrate assemblages and leading to the high variability seen in many such datasets[15], [35].

A reduction in macroinvertebrate FD examined in this study translates into reduced community-level variation in foraging mechanisms and locomotion/substrate relations, which is likely to lead to significant alterations in food web structure and energy flow in those coastal wetland systems. Reduced FD is also likely to translate into decreased asynchrony of taxa responses to environmental perturbations (because more functionally similar taxa would have more synchronized responses to changes in resource abundance), which in turn has been proposed to be one of the key mechanisms reducing ecosystem stability in a theoretical modeling study [39].

The greater relative importance of development vs. agricultural stress is not surprising considering previously reported macroinvertebrate community thresholds at very low levels of development [40]–[44]. Similarly, a previous study found that development in the watershed was the best predictor of several macroinvertebrate metrics, including *Aeshna* abundance [37]. However, further study is needed to identify the more proximal factors driving this relationship, specifically in littoral systems. For instance, decline in longer-lived organisms could be mediated through the greater cumulative effects of development-associated stressors such as point-source pollution, in particular organic pollution, on these organisms, or through the destruction of biogenic (macrophyte) complexity, which has been shown to support greater abundance of several types of predators [45] and likely sustains more stable, and thereby more diverse, predator-prey populations [46]. It has proven difficult to elucidate repeatable patterns in these coastal wetlands, possibly due to their highly variable hydrology [35], [47], e.g. half a meter or more changes in water level on diel and annual basis due to fetch and climate change. Despite this, there is considerable interest in finding community metrics responsive to anthropogenic stress, and several such metrics have been found including Ephemeroptera-Trichoptera-Odonata richness-based metrics [48].

The observation of shift in functional community composition towards greater relative abundance of burrowers and filterers is consistent with other studies and has been previously related to increased sedimentation in streams [49]. Yet flashier hydrology and increased siltation, which are cited as the most common causes of anthropogenic changes in stream assemblages [1], are less likely to be the same factors responsible for observed changes in littoral assemblages, and lake-specific environmental variables responsible for these effects need to be investigated. Several functional groups (clingers, burrowers and insect filter-gatherers) were previously demonstrated to be affected by the extent of anthropogenic development in this system; however, responses were complex and dependent on several predictor variables as well as land use predictors' buffer sizes [37].

Trait responses are generally less frequently reported in aquatic studies than trends in diversity and abundance [50]. Effects of watershed land use on FD have garnered even less attention, particularly in non-stream ecosystems. For lotic macroinvertebrates, it has been shown that trait type changed [51] and trait diversity decreased [52] with increasing agricultural land use. The latter study demonstrated that this effect was detectable both at the watershed scale and local patch scale, and was related to increasing sedimentation [52]. Significant effects on these important functional variables (e.g., trait diversity) emphasize the need to further investigate functional responses to anthropogenic stress in lentic macroinvertebrates, the mechanisms potentially underlying these responses and the local factors that may mitigate those effects.

Considering additional local factors as well as integrating traits with explicit consideration of trait linkages [53] may increase the predictive capability of future littoral trait models. However, what is of greater interest for future studies is that this uncertainty potentially reflects trait responses to smaller-scale habitat factors, which, if amendable to manipulation or restoration, could be used to ameliorate effects of watershed-scale land use.

In summary, we observed statistically and biologically significant reduction in macroinvertebrate FD and abundance of longer-lived taxa, and noticeable differences in functional community structure associated with increasing proportion of development in the watershed. These findings, along with the previously-observed threshold changes in macroinvertebrate community composition [44], show that lentic fauna exhibit significant functional changes associated with greater levels of watershed-scale anthropogenic stress.

ACKNOWLEDGMENTS

We thank field and lab personnel from NRRI-University of Minnesota Duluth, and University of Windsor for their assistance in collecting and processing samples and data used in this manuscript, and Terry Brown for creating the maps. We are grateful to Arthur Compin and an anonymous reviewer for their constructive comments. This is contribution number 572 of the Center for Water and the Environment, Natural Resources Research Institute, University of Minnesota Duluth.

REFERENCES

1. Allan JD (2004) Landscapes and riverscapes: the influence of land use on stream ecosystems. Annu Rev Ecol Evol Syst 35: 257–284. doi: 10.1146/annurev.ecolsys.35.120202.110122
2. Heino J (2010) Are indicator groups and cross-taxon congruence useful for predicting biodiversity in aquatic ecosystems? Ecol Indic 10: 112–117. doi: 10.1016/j.ecolind.2009.04.013
3. Statzner B, Bêche L (2010) Can biological invertebrate traits resolve effects of multiple stressors on running water ecosystems? Freshw Biol 55: 80–119. doi: 10.1111/j.1365-2427.2009.02369.x

4. Díaz S, Cabido M (2001) Vive la différence: plant functional diversity matters to ecosystem processes. Trends Ecol Evol 16: 644–655. doi: 10.1016/s0169-5347(01)02283-2
5. Folke C, Carpenter SR, Walker BH, Scheffer M, Elmqvist T, et al. (2004) Regime shifts, resilience and biodiversity in ecosystem management. Annu Rev Ecol Evol Syst 35: 557–581. doi: 10.1146/annurev.ecolsys.35.021103.105711
6. Dunne JA, Brose U, Williams RJ, Martinez ND (2005) Modeling food web dynamics: complexity-stability implications. In: Belgrano A, Scharler UM, Dunne J, Ulanowics RE, editors. Aquatic food webs: an ecosystem approach. Oxford University Press, New York. 117–129.
7. Hooper DU, Solan M, Symstad A, Diaz S, Gessner MO, et al.. (2002) Species diversity, functional diversity and ecosystem functioning. Chapter 17 in Loreau M, Naeem S, Inchausti P, editors. Biodiversity and Ecosystem Functioning: a Current Synthesis. Oxford University Press. 147–154.
8. McGill BJ, Enquist BJ, Weiher E, Westoby M (2006) Rebuilding community ecology from functional traits. Trends Ecol Evol 21: 178–185. doi: 10.1016/j.tree.2006.02.002
9. Loreau M (1998) Biodiversity and ecosystem functioning: A mechanistic model. Proc Natl Acad Sci USA 95: 5632–5636. doi: 10.1073/pnas.95.10.5632
10. Dolédec S, Phillips N, Scarsbrook M, Riley RH, Townsend CR (2006) Comparison of structural and functional approaches to determining landuse effects on grassland stream invertebrate communities. J North Am Benthol Soc 25: 44–60. doi: 10.1899/0887-3593(2006)25[44:cosafa]2.0.co;2
11. Covich AP, Palmer MA, Crowl TA (1999) The role of benthic invertebrate species in freshwater ecosystems. BioScience 49: 119–127. doi: 10.2307/1313537
12. Cardinale BJ, Palmer MA, Collins SL (2002) Species diversity increases ecosystem functioning through interspecific facilitation. Nature 415: 426–429. doi: 10.1038/415426a
13. Merritt RW, Cummins KW, Berg MB, editors (2008) An Introduction to the Aquatic Insects of North America: Fourth Edition. Kendall/Hunt Publishing Company. 1214 p.

14. Huryn AD, Wallace JB, Anderson NH (2008) Habitat, life history, secondary production and behavioral adaptations of aquatic insects. Chapter 5 in: Merritt RW, Cummins KW, Berg MB. An Introduction to the Aquatic Insects of North America. Kendall/Hunt Publishing Company. 55–104.
15. Batzer DP (2013) The seemingly intractable ecological responses of invertebrates in North American wetlands: a review. Wetlands 33: 1–15. doi: 10.1007/s13157-012-0360-2
16. Niemi GJ, Kelly JR, Danz NP (2007) Environmental indicators for the coastal region of the North American Great Lakes: Introduction and prospectus. J Great Lakes Res 33: 1–12. doi: 10.3394/0380-1330(2007)33[1:eiftcr]2.0.co;2
17. Danz NP, Regal RR, Niemi GJ, Brady VJ, Hollenhorst TP, et al. (2005) Environmentally stratified sampling design for the development of Great Lakes environmental indicators. Environ Monit Assess 102: 41–65. doi: 10.1007/s10661-005-1594-8
18. Hollenhorst TP, Johnson LB, Ciborowski J (2011) Monitoring land cover change in the Lake Superior basin. Aquat Ecosyst Health Manag 14: 433–442. doi: 10.1080/14634988.2011.628242
19. Merritt RW, Cummins KW, editors (1995) An introduction to the aquatic insects of North America. Third edition. Kendall/Hunt Publishing Company. 862 p.
20. Thorp JH, Covich AP, editors (2001) Ecology and classification of North American freshwater invertebrates. Second edition. Academic Press. 1056 p.
21. Courtney GW, Merritt RW (2008) Aquatic Diptera Part One. Larvae of aquatic Diptera. Chapter 22 in: Merritt RW, Cummins KW, Berg MB. An Introduction to the Aquatic Insects of North America. Kendall/Hunt Publishing Company. 687–722.
22. Morse JC, Holzenthal RW (2008) Trichoptera Genera. Chapter 18 in: Merritt RW, Cummins KW, Berg MB. An Introduction to the Aquatic Insects of North America. Kendall/Hunt Publishing Company. 481–552.
23. Polhemus JT (2008) Aquatic and semiaquatic Hemiptera. Chapter 15 in: Merritt RW, Cummins KW, Berg MB. An Introduction to the Aquatic Insects of North America. Kendall/Hunt Publishing Company. 385–423.

24. Solis MA (2008) Aquatic and semiaquatic Lepidoptera. Chapter 19 in: Merritt RW, Cummins KW, Berg MB. An Introduction to the Aquatic Insects of North America. Kendall/Hunt Publishing Company. 553–569.
25. Tennessen KJ (2008) Odonata. Chapter 12 in: Merritt RW, Cummins KW, Berg MB. An Introduction to the Aquatic Insects of North America. Kendall/Hunt Publishing Company. 237–294.
26. Waltz RD, Burian SK (2008) Ephemeroptera. Chapter 11 in: Merritt RW, Cummins KW, Berg MB. An Introduction to the Aquatic Insects of North America. Kendall/Hunt Publishing Company. 181–236.
27. White DS, Roughly RE (2008) Aquatic Coleoptera. Chapter 20 in: Merritt RW, Cummins KW, Berg MB. An Introduction to the Aquatic Insects of North America. Kendall/Hunt Publishing Company. 571–671.
28. Rao CR (1982) Diversity and dissimilarity coefficients – a unified approach. Theor Popul Biol 21: 24–43. doi: 10.1016/0040-5809(82)90004-1
29. Botta-Dukát Z (2005) Rao's quadratic entropy as a measure of functional diversity based on multiple traits. J Veg Sci 16: 533–540. doi: 10.1658/1100-9233(2005)16[533:rqeaam]2.0.co;2
30. Laliberté E, Legendre P (2010) A distance-based framework for measuring functional diversity from multiple traits. Ecology 91: 299–305. doi: 10.1890/08-2244.1
31. Lindeman RH, Merenda PF, Gold RZ (1980) Introduction to Bivariate and Multivariate Analysis, Glenview, IL: Scott, Foresman. 444 p.
32. Grömping U (2006) Relative importance for linear regression in R: the package relaimpo. J Stat Softw 17: 1–27.
33. Watson D (1994) Nngridr-An Implementation of Natural Neighbor Interpolation. D. Watson. 170 p.
34. Hunter JD (2007) Matplotlib: A 2D graphics environment. Comput Sci Eng 9: 90–95. doi: 10.1109/mcse.2007.55
35. Poff NL (1997) Landscape filters and species traits: towards mechanistic understanding and prediction in stream ecology. J North Am Benthol Soc 16: 391–409. doi: 10.2307/1468026

36. Weigel BM, Wang L, Rasmussen PW, Butcher JT, Stewart PM, et al. (2003) Relative influence of variables at multiple spatial scales on stream macroinvertebrates in the Northern Lakes and Forest ecoregion, U.S.A. Freshw Biol. 48: 1440–1461. doi: 10.1046/j.1365-2427.2003.01076.x
37. Brazner JC, Danz NP, Niemi GJ, Regal RR, Hollenhorst T, et al. (2007) Responsiveness of Great Lakes wetland indicators to human disturbances at multiple spatial scales: a multi-assemblage assessment. J Great Lakes Res 33: 42–66. doi: 10.3394/0380-1330(2007)33[42:roglwi]2.0.co;2
38. Compin A, Céréghino R (2007) Spatial patterns of macroinvertebrate functional feeding groups in streams in relation to physical variables and land-cover in Southwestern France. Landsc Ecol 22: 1215–1225. doi: 10.1007/s10980-007-9101-y
39. Loreau M, de Mazancourt C (2013) Biodiversity and ecosystem stability: a synthesis of underlying mechanisms. Ecol Lett 16: 106–115. doi: 10.1111/ele.12073
40. King RS, Baker ME (2010) Considerations for analyzing ecological community thresholds in response to anthropogenic environmental gradients. J North Am Benthol Soc 29: 998–1008. doi: 10.1899/09-144.1
41. Hilderbrand RH, Utz RM, Stranko SA, Raesly RL (2010) Applying thresholds to forecast potential biodiversity loss from human development. J North Am Benthol Soc 29: 1009–1016. doi: 10.1899/09-138.1
42. Utz RM, Hilderbrand RH (2011) Interregional variation in urbanization-induced geomorphic change and macroinvertebrate habitat colonization in headwater streams. J North Am Benthol Soc 30: 25–37. doi: 10.1899/10-007.1
43. Kail J, Arle J, Jähnig SC (2012) Limiting factors and thresholds for macroinvertebrate assemblages in European rivers: Empirical evidence from three datasets on water quality, catchment urbanization, and river restoration. Ecol Indic 18: 63–72. doi: 10.1016/j.ecolind.2011.09.038
44. Kovalenko KE, Brady VJ, Brown TN, Ciborowski JJH, Danz NP, et al. (2014) Congruence of community thresholds in response to anthropogenic stressors in Great Lakes coastal wetlands. Freshw Sci (In press).

45. Heino J (2008) Patterns of functional biodiversity and function–environment relationships in lake littoral macroinvertebrates. Limnol Oceanogr 53: 1446–1455. doi: 10.4319/lo.2008.53.4.1446
46. Kovalenko KE, Thomaz SM, Warfe DM (2012) Habitat complexity: approaches and future directions. Editorial review. Hydrobiologia 685: 1–17. doi: 10.1007/s10750-011-0974-z
47. Gathman JP, Burton TM (2011) A Great Lakes coastal wetland invertebrate community gradient: relative influence of flooding regime and vegetation zonation. Wetlands 31: 329–341. doi: 10.1007/s13157-010-0140-9
48. Uzarski DG, Burton TM, Genet JA (2004) Validation and performance of an invertebrate index of biotic integrity for Lakes Huron and Michigan fringing wetlands during a period of lake level decline. Aquat Ecosyst Health Manag 7: 269–288. doi: 10.1080/14634980490461498
49. Larsen S, Pace G, Ormerod SJ (2011) Experimental effects of sediment deposition on the structure and function of macroinvertebrate assemblages in temperate streams. River Res Appl 27: 257–267. doi: 10.1002/rra.1361
50. Stendera S, Adrian R, Bonada N, Cañedo-Argüelles M, Hugueny B, et al. (2012) Drivers and stressors of freshwater biodiversity patterns across different ecosystems and scales: a review. Hydrobiologia 696: 1–28. doi: 10.1007/s10750-012-1183-0
51. Richards C, Haro RJ, Johnson LB, Host GE (1997) Catchment and reach-scale properties as indicators of macroinvertebrate species traits. Freshw Biol 37: 219–230. doi: 10.1046/j.1365-2427.1997.d01-540.x
52. Larsen S, Ormerod SJ (2010) Combined effects of habitat modification on trait composition and species nestedness in river invertebrates. Biol Conserv 143: 2638–2646. doi: 10.1016/j.biocon.2010.07.006
53. Verberk WCEP, van Noordwijk CGE, Hildrew AG (2013) Delivering on a promise: integrating species traits to transform descriptive community ecology into a predictive science. Freshw Sci 32: 531–547. doi: 10.1899/12-092.1

Chapter 5

The Effects of Spatial Scale on Breakdown of Leaves in a Tropical Watershed

Renan S. Rezende[1,2]*, Mauricio M. Petrucio[1], José F. Gonçalves, Jr.[2]

[1]Graduate Program in Ecology of Department of Ecology and Zoology, Institute of Biological Sciences, Federal University of Santa Catarina, Florianópolis, Santa Catarina, Brazil,

[2]Department of Ecology, Institute of Biology, University of Brasília, Brasília, Federal District, Brazil

ABSTRACT

The objective was to assess the effects of natural variation in the physical structure of the environment on biological communities and on the

processing of *Eucalyptus cloeziana* and*Inga laurina* and to identify the controlling factors at different scales along stream order gradients. The study area consisted of 14 sampling sites distributed within a tropical watershed (1st, 2nd, 3rd and 4th order streams replicated in 4 sub-basins). Our samples consisted of 3 g of leaves of *E. cloeziana* (high-quality) and *I. laurina* (low-quality) placed in 252 bags with 10mm mesh (measured by the chemical composition of the detritus). Four samples of each leaf type were collected periodically (three times) over a period of 75–125 days and washed on a sieve to separate the invertebrates. A series of leaf disks were cut to determine ash-free dry mass, polyphenol, lignin, cellulose, total microbial biomass and fungal biomass, and the remaining material was oven-dried to determine the dry weight. We performed analyses within and between spatial scales (regional and local) to assess which watershed scale was the more import determinant of the leaf breakdown rate (*k*). The microbial and shredder were most influenced at the local scale (stream order). Shredders were influenced by microorganisms, with stronger interactions between them than were found to drive the *k* at the local scale. Moreover, differences in the overall *k* and abiotic variables were more strongly influenced at the regional scale (sub-basin), showing that the study scale alters the response of the studied variables. We found higher *k* values at higher values of water velocity, dissolved oxygen and temperature, all of which accelerate biological metabolism in response to variations on the regional scale. Watersheds with warmer microclimates and streams with higher nutrient levels and oxygen could be accelerating the ecosystem metabolism, independent of the detritus quality.

INTRODUCTION

The characteristics of lotic ecosystems show natural patterns along an upstream-downstream gradient due to variations in geomorphology and topography in the watershed [1], [2]. Natural changes along this gradient (upstream to downstream) include an increase in the dimensions of the stream (width), changes (increases or decreases) in the velocity of the water, and openings in the riparian canopy that allow greater light penetration [3], [4]. Greater luminosity increases the temperature and accelerates photosynthetic production and autotrophic metabolism [3]. Along this gradient, therefore, the relative abundance of micro-organisms increases but that of invertebrates decreases (primarily shredders of organic matter), decreases from the headwaters to the downstream sections [1]. These natural changes also modify the energy input and cycling of organic matter in space and time [1], [5]. Senescent leaves are an important source of nutrients and food resources in heterotrophic metabolic environments, primarily in headwaters and small streams [6], [7]. However, this material is mineralized and available for use by primary producers and other trophic levels after its decomposition. As a result, leaf breakdown is a key process in lotic ecosystems [6], [8]. Leaf breakdown can be influenced by many factors, such as physical and chemical variables (associated with water and detritus) and the activities of communities of decomposers (micro-organisms and aquatic invertebrates) [6]–[13].

The study of leaf breakdown at the scale of a watershed allows us to observe emerging patterns and identify certain factors that structure the ecosystems at different scales [14],[15]. It is evident that a series of successively smaller and nested geomorphologic units can have various patterns and structures depending on the scale that is being analyzed [14], [16]. These patterns and structures can be observed in riffles and pools within continuous stretches, which are nested within large rivers that make up a watershed [15], [17], [18]. Studies that address only one scale are subject to problems because certain variables are measured directly in small areas or across short time intervals, whereas few can be measured at fine resolution over large areas [17], [18]. In addition, changes at smaller scales are not maintained at larger scales [14]. Therefore, the issue is that unless patterns are consistent at all scales, the findings at one scale cannot be extrapolated to yield accurate predictions at other scales. Accordingly, tests at multiple scales are needed for confident extrapolation. From this perspective, the evaluation of leaf breakdown at different scales enables the development of an integrated vision of the landscape during this important ecological process [19], [20].

The streams in a watershed can be considered within a hierarchical framework that presents organized view of spatial and temporal variations among and within stream systems along the "riverscapes" [21]–[23]. Therefore, several studies have examined leaf processing at large spatial scales across biomes [24], [25], latitudinal gradients [24], and altitudinal gradients [2]and influences of land use [26], [27]. Moreover, several recurrent topics emerge from considerations of several spatial scales. These topics include the relative importance of fungi and invertebrates [28], the use of bioindicators [20] and the hierarchical nature of lotic ecosystems [15]. Studies assessing allochthonous leaf breakdown at a watershed scale are rare worldwide, but they have been performed in temperate systems [15], [20], [23], [29].

In tropical streams, individual riffles or short stream reaches continue to be the most frequent sites for studies of leaf processing based on the traditional conceptual model [30], [31]. Several factors are known to cause variation in the rates of processing within and among tropical stream reaches [13], [32]. These factors include the effects of species mixing [33],[34], litter quality [35], [36], micro-organism communities [37], [38], invertebrate communities[39]–[41], detritivores and shredders [10], [42] and seasonal effects [8], [43], [44]. However, systematic assessments of variability in allochthonous leaf breakdown rates across multiple spatial scales using the watershed as the sampling unit have not been performed in tropical stream systems. This study could help to answer important questions, such as "how does spatial structure influence ecosystem function and how do we integrate within and between spatial scales to assess function", suggested by Sutherland et al. [45] as one of 100 fundamental ecological questions.

Based on the premise that leaf breakdown is the result of the activity of decomposer organisms and the physical and chemical processes occurring in the stream water, which vary along the scale investigated in the study scale [6], [12], [13], the following hypotheses were tested in this study: (i) natural differences in the physical nature of the stream (increasing canopy opening, water velocity, temperature and nutrient concentrations) accelerate biological metabolism and leaf breakdown from upstream to downstream; (ii) shredders decrease and micro-organisms increase in importance from upstream to downstream; and (iii) differences in the overall *k* values will be more clearly understandable (strongly explained) at an increased spatial scale. The objective of the study was to assess the natural effects of variation in the physical environment on biological communities and the leaf breakdown rates of *Eucalyptus cloeziana F. Muell* and *Inga laurina Sw. Willd* and to identify the controlling factors at different scales along the stream order gradient.

METHODS

The Study System

The study area consisted of 14 sampling sites distributed along the Gama-Cabeça do Veado watershed, a part of the Federal District in west central Brazil, comprising 1st, 2nd, 3rd and 4th order streams replicated in 4 sub-basins (Figure 1). The area includes important waterways that form the Paraná basin, a part of the Cerrado biome (Brazilian Savannah). The climate is tropical and has distinct rainy-hot (October to April) and cold-dry seasons (May to September). The mean annual temperature is 20°C, and the altitude varies between 1025 and 1150 m above sea level. The study area included three conservation units protecting the entire watershed and all sampling sites studied (Ecological Station of the University of Brasília, Ecological Reserve of IBGE and Ecological Station of the Botanical Gardens of Brasília). The study was approved by Ministry of Environment of Brazil through the System of Information and Authorization on Biodiversity (SISBIO) for activities with scientific purpose (code: 39629-1), and also was approved by the Scientific and Technical Council of the Ecological Station of the University of Brasília (code: 05–12), IBGE Ecological Reserve (code: 54 PC - PAD 1) and Botanical Gardens of Brasília (code: 13/2011).

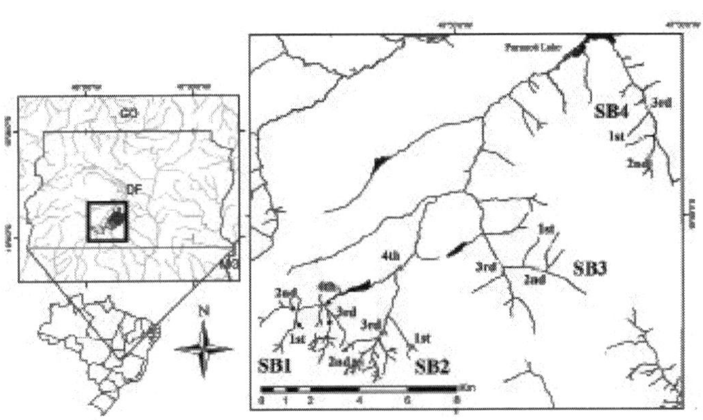

Figure 1. Sampling sites. Geographic location of the sampling sites consisting of streams of the 1st, 2nd, 3rd and 4th orders replicated in 4 sub-basins in the Gama-Cabeça de Veado watershed, Federal District (Brazil).

Procedures

The experiment was conducted from June through October 2011 (the winter season), a period of zero rainfall, resulting in a high level of homogeneity in the physical and chemical features of the environment. Based on their chemical composition, we selected leaves from two species for use as detritus. The leaves of an exotic species (*Eucalyptus cloeziana F. Muell*) were used to represent high-quality detritus. The planting of *Eucalyptus* monocultures in place of the native vegetation has potential repercussions for stream basins. The area in which the study was conducted does not contain *Eucalyptus* plantations. However, the substitution of this monoculture for native vegetation has occurred in neighboring basins, where this monoculture has expanded, as it has throughout Brazil, e.g., to supply charcoal for steelmaking and pulp for papermaking [35]. The low-quality detritus from a native species (abundant in riparian vegetation; *Inga laurina Sw. Willd*) was also used in the study, as this species best represents the plants of the Cerrado.

The leave The two types of leaves were chemically characterized by the mean values of total polyphenols (22.80 ± 2.5; 18.29 ± 1.8 mg/g^{-1}), total tannic acids (0.003 ± 0.0002; 0.002 ± 0.0004 mg/g^{-1}), lignin (42.61 ± 0.7; 45.94 ± 0.5%), cellulose (24.69 ± 1.5; 37.39 ± 1.2%), hardness (0.17 ± 0.1; 0.6 ± 0.3 cm/g^{-1}), nitrogen (13.16 ± 1.3; 16.41 ± 1.0 g/kg^{-1}) and phosphorus (0.46 ± 0.05; 0.53 ± 0.07 g/kg^{-1}) in *E. cloeziana* and *I. laurina*, respectively. The breakdown rates for these two leaf types (collected in nets 1 m^2 in area placed 1.5 m from the ground) were measured individually by the loss of weight of 3 g (± 0.1 g dry weight) of leaves, correcting for the initial humidity and transport loss [46], incubated in litter bags (15 × 15 cm, 10 mm mesh size).

In total, 252 litter bags were placed at a depth of 0.3 m in pool areas at the 14 sampling sites in 1st, 2nd, 3rd and 4th order streams [47] in the 4 sub-basins (Gama – SB1, Capetinga – SB2, Taquaras – SB3 and Cabeça de Veado – SB4 sub-basins). The removal of the *E. cloeziana*and *I. laurina* leaves occurred initially after 10 days of incubation, which corresponds to the expected time required for leaching and initiating microbial colonization [46]. After this sampling, the principal leaf breakdown rate (k) was used to estimate the next sampling time for each detritus type (at approximately 75% of the remaining mass), which was determined to be 40 days for *E. cloeziana* and 85 days for *I. laurina*. This stage of decomposition occurs when microbial activity is high and the invertebrate community is already established. Subsequently, the *k* value was used a second time to perform corrections and acquire new values to obtain approximately 50% of the remaining mass for each detritus type. This sampling time was determined to be 75 days for *E. cloeziana* and 125 days for *I. laurina*. After this period, the community established during the ecological process of degradative succession is affected by the reduction in the available detritus (additional information about k, see also Chapter 6 of[46]).

The sampling times were calculated by dividing the initial weight (W0) by the estimated value of *k*. This calculation yields the time for the total course of leaf processing (TLP, days). From the equation *W0/k = TLP*, we can calculate how many days will be required to reach a desired percentage of the initial weight (Wt). The first sample was collected after 10 days of incubation for both species, so that TLP for 10 days/0.25 = day on which Wt = 75%. The next sample was collected after 40 days for *E. cloeziana* and after 75 days for *I. laurina*, so that TLP 40/75 days/0.5 = day by which Wt = 50%. The above procedure was performed for each sample site (based on the mean value) and type of detritus. However, it was not possible to determine the final value for *I. laurina* because the dry season ended after 120 days, before 50% of the mass had been lost. Measurements after the end of the dry season would not have been meaningful because variations in rainfall and associated variations in other physical and chemical conditions would have influenced the results.

On removal from the streams, the litter bags were placed individually into insulated plastic bags and transported in thermal containers (± 4°C) to the laboratory. Temperature, electrical conductivity, pH, dissolved oxygen and water turbidity were obtained *in situ* with a multi-analyzer measured each time leaf bags were removed. The depth and average speed of the right, left and central portions of the watercourse were measured with a flow-meter, and the instantaneous discharge of water was then calculated. We collected 1 L of water to determine the nitrate [48], ammonia [49] and orthophosphate [50] concentrations. The canopy openings were quantified using hemispherical photographs taken with a digital camera equipped with a fish-eye lens. These photographs were later analyzed using Gap Light Analyzer software (2.0). The leaves were washed with tap water in a 120 μm mesh sieve. The invertebrates retained on the sieve were preserved in 70% alcohol for later identification and counting [51], [52]. The numbers of *taxa* and individuals were calculated for the aquatic invertebrate community, and biomass was obtained by desiccation at 60°C for 72 h. The invertebrates were classified into five feeding categories [51]–[53]: gathering-collectors (G–C), filtering-collectors (F–C), shredders (Sh), scrapers (Sc) and predators (P).

Five leaves from each sample were randomly collected, and three disks (1.2 cm diameter) were extracted from each leaf, resulting in three five-disk sets. One set was used to determine the remaining ash-free dry mass (AFDM; calculated after incineration in a muffle furnace at 550°C for 4 h), and the other sets were used to assess the ergosterol and ATP concentrations. The remaining material was oven-dried at 60°C for 72 h to determine its dry weight. The leaf breakdown rates (*k*) were calculated using the negative exponential model of percent mass lost over time ($W_{t=W0}e^{-kt}$; Wt = remaining weight; W0 = initial weight; –k = decay rate; t = time). After the leaves had been dried and weighed, they were pulverized for further analysis of the total polyphenol and tannic acid concentration [54], lignin and cellulose contents [55] and the resistance of leaves to rupture (hardness of intact leaves [46]). Values for total nitrogen were obtained using a CHN basic analyzer (Carlo Erba 1500 for WI; Thermo Electron Corp. Milan, Italy), and values for total phosphorus were obtained using the ascorbic acid method after acid digestion. The total micro-organism biomass was measured by quantifying ATP [56]. The biomass of aquatic Hyphomycetes was evaluated by quantifying ergosterol, a lipid exclusive to fungal membranes in this community [57].

Data Analysis

An analysis of variance (function lm, package stats for R version 2.12.1; [58]) was used to analyze the physical and chemical parameters of the water (temperature, electrical conductivity, pH, dissolved oxygen, turbidity, nitrites, nitrates, orthophosphates and mean velocity) and the structure of stream stretches (instantaneous discharge of the stream and canopy openings in riparian vegetation) as dependent variables, using two categorical factors, namely, sub-basins and stream order. Stream order was also used as a co-variate (continuous variable). We also used the leaf mass remaining, invertebrate communities (number of *taxa*, density and biomass), the relative abundance of functional trophic groups of invertebrates (gathering-collectors, filtering-collectors, shredders, scrapers and predators) and microbial biomass (ATP and ergosterol) as dependent variables against the same two categorical factors. Stream order was also used as a co-variate (continuous variable). This procedure was performed similarly for both types of detritus. All models used a Gaussian distribution (link = log; test = F). We used an analysis of contrasts to discriminate among categorical variables. The normality of the data was tested using a Kolmogorov-Smirnov test, the homogeneity of variance was determined with a Levene test, and the data were transformed whenever necessary with the Naperian logarithm (ln) to obtain the best fit [58].

RESULTS

Abiotic Variables

The values of instantaneous discharge, electrical conductivity and nitrates were the highest in the 3rd and 4th order streams. In contrast, the 1st order streams had the highest values for temperature, canopy opening and nitrite concentrations in the water, and the water velocity was the lowest. Dissolved oxygen, pH, turbidity and orthophosphates did not differ among the stream orders (analysis of contrasts, $p < 0.05$; Table S1, Table 1, Figure 2). In sub-basin 2 (SB2), we observed high electrical conductivity, high nitrite and nitrate concentrations and low water temperatures. The values for canopy opening and water velocity were highest in sub-basins 1 (SB1) and 4 (SB4), respectively. Dissolved oxygen and orthophosphates were highest in sub-basin 3 (SB3). Instantaneous discharge, pH and turbidity did not differ among the sub-basins (analysis of contrasts, $p < 0.05$; Table S1, Figure 2). We observed that the higher percentages of sums of squares and variance in instantaneous discharge and water velocity could be explained by differences in the stream order. However, dissolved oxygen, electrical conductivity, temperature, pH, turbidity, canopy opening, nitrates, nitrites and orthophosphates exhibited a high level of variance among the sub-basins (Table 1).

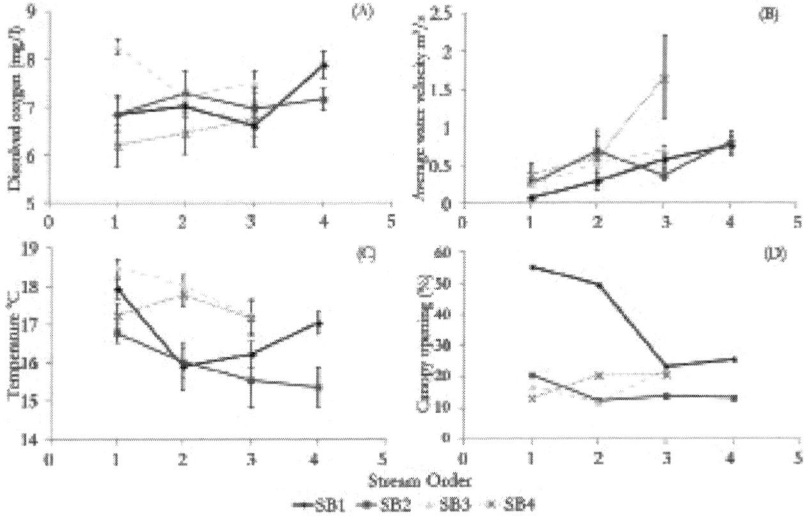

Figure 2. Abiotic variables.

Mean values and standard errors for dissolved oxygen (A), water velocity (B), water temperature (C) and canopy opening (D) for the stream orders and among sub-basins.

Table 1. Degrees of freedom (DF), residuals, sums of squares (%), F tests and analyses of variance and contrasts (AC) for dissolved oxygen, electrical conductivity, water temperature, pH, turbidity, water velocity, canopy openness and nitrites, nitrates and orthophosphates in sub-basins and along the stream orders in the Gama-Cabeça de Veado Basin

s		Df	Sum Sq (%)	F value	Pr(>F)	AC
Instantaneous discharge	Sub-Basin	3	5.71	1.97	0.127	
	Order	1	31.56	32.71	<0.001	1st = 2nd < 3rd = 4th
	Residuals	65	62.72			
Dissolved oxygen	Sub-Basin	3	11.34	2.80	0.047	SB4 = SB2 = SB1 <SB3
	Order	1	0.87	0.65	0.425	

	Residuals	65	87.78			
Electrical conductivity	Sub-Basin	3	42.63	27.22	<0.001	SB4 = SB1 = SB3 <SB2
	Order	1	23.42	44.86	<0.001	1st = 2nd<3rd = 4th
	Residuals	65	33.93			
Temperature	Sub-Basin	3	22.39	6.66	<0.001	SB2 <SB1 = SB4 = SB3
	Olt	1	0.515	4.62	0.035	4th = 3rd=2nd < 1st
	Residuals	65	72.55			
pH	Sub-Basin	3	5.44	1.25	0.298	
	Order	1	0.47	0.33	0.567	
	Residuals	65	94.07			
Turbidity	Sub-Basin	3	2.66	0.62	0.605	
	Order	1	4.23	2.96	0.090	
	Residuals	65	93.10			
Average water velocity	Sub-Basin	3	10.44	3.79	0.014	SB1 = SB3 = SB2 <SB4
	Order	1	29.77	32.37	<0.001	1st < 2nd = 4th=3rd
	Residuals	65	59.77			
Canopy opening	Sub-Basin	3	59.83	41.13	<0.001	SB2 = SB3 = SB4 <SB1
	Order	1	8.64	17.83	<0.001	4th =3rd = 2nd<1st
	Residuals	65	31.52			

Nitate	Sub-Basin	3	42.56	32..68	<0.001	SB3 = SB4 = SB1 <SB2
	Order	1	29.22	67.32	<0.001	$1^{st} < 2^{nd} = 3^{rd} < 4^{th}$
	Remluds	65	28.21			
Nitrite	Sub-Basin	3	35.50	23.86	<0.001	SB4 = SB3 <SB1 <SB2
	Order	1	32.25	65.02	<0.001	$4^{th} < 2^{nd} = 3^{rd} < 1^{st}$
	Residuals	65	32.24			
Orthophosphate	Sub Basin	3	11.62	2.87	0.043	SB3< SB1<SB4s <SB2
	Order	1	0.54	0.40	0.527	
	Residuals	65	87.83			

Leaf Breakdown Rates

The leaf breakdown rates (*k*) were the highest in the 2nd order streams, with values of –0.0083 and –0.0022 for *E. cloeziana* and *I. laurina,* respectively, followed by the 3rd order stream sections, with values of –0.0071 and –0.0022. We also observed higher *k* values in 1st order streams (–0.0053 and –0.0015) than in 4th order streams (–0.0051 and –0.0018) for *E. cloeziana* and *I. laurina,* respectively. However, the remaining mass did not differ among stream orders for either of the detritus types (Figure S1, Table 2, Figure 3A and 4A). The highest *k* values were observed in SB4 (–0.0105 and –0.0030), followed by SB3 (–0.0088 and –0.0022 for *E. cloeziana* and *I. laurina,* respectively). For *E. cloeziana,* the values were –0.0062 and –0.0049, whereas *I. laurina* exhibited values of –0.0018 and –0.0016 (for SB1 and SB2, respectively). The remaining mass showed the lowest values in SB4 and SB3 among the sub-basins studied. The variance in the remaining mass was higher (by sums of squares) and also explained the variations in the sub-basins (Figure S1, Table 2, Figure 3A and 4A).

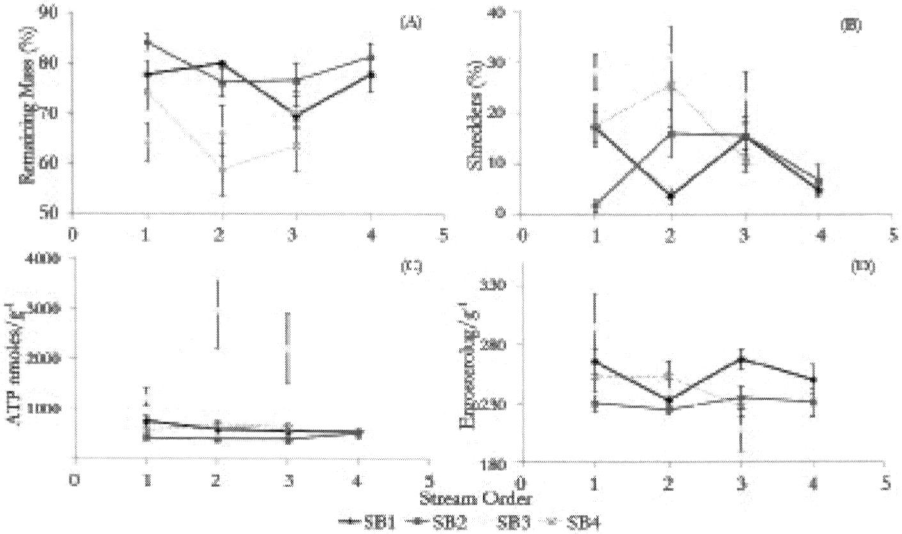

Figure 3. Leaf breakdown process in E. cloeziana. Mean values and standard errors for the remaining mass (A), shredder abundance (B), total microbial biomass (ATP; C) and fungal hyphomycetal biomass (Ergosterol; D) for the stream orders and sub-basins for *E. cloeziana*.

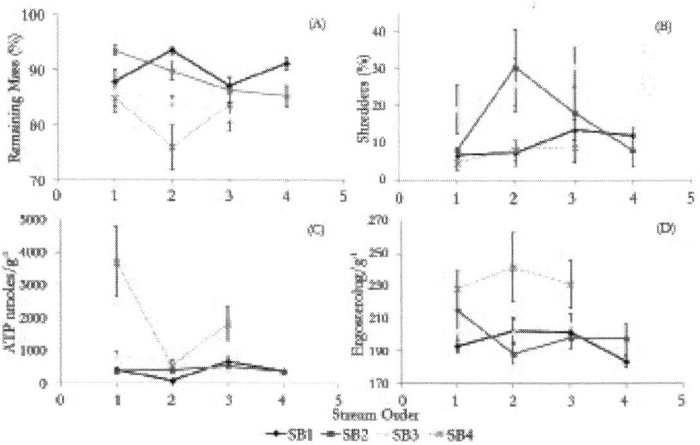

Figure 4. Leaf breakdown process in I. laurina. Mean values and standard errors for the remaining mass (A), shredder abundance (B), total microbial biomass (ATP; C) and fungal hyphomycetal biomass (Ergosterol; D) for the stream orders and sub-basins for *I. laurina*.

Table 2. Degrees of freedom (DF), residuals, sums of squares (%), F test and analyses of variance and contrasts (AC) for the remaining mass, density, richness and biomass of invertebrates, functional trophic groups of invertebrates (predators, shredders, gatherer-collectors, filtering-collectors and scrapers), ATP and ergosterol in sub-basins and along the stream orders in the Gama-Cabeça de Veado Basin for E. cloeziana and I. laurina

E.cloeziana						I. laurina					
		Df	Sum Sq (%)	F value	Pr(>F)	AC	Df	Sum Sq (%)	F value	Pr(>F)	**AC**
Remaining mass	Sub-Basin	3	15.05	6.51	<0.001	SB4=SB3<SB1=SB2	3	11.01	4.72	0.004	SB4=SB3 <SB2=SB1
	Order	1	0.18	0.24	0.628		1	1.95	2.51	0.116	
	Residuals	110	84.76				112	87.04			
Invertebrate Density	Sub-Basin	3	0.63	0.25	0.864		3	9.29	4.14	0.008	SB2=SB1 <SB3=SB4
	Order	1	5.49	6.43	0.013	$2^{nd}=1^{st}=3^{rd}<4^{th}$	1	6.88	9.19	0.003	$2^{nd}=1^{st}=3^{rd}<4^{th}$
	Residuals	110	93.88				112	83.83			
Invertebrate Richness	Sub-Basin	3	12.42	5.65	0.001	SB1=SB2<SB4=SB3	3	25.53	15.24	<0.001	SB2<SB1 =SB3=SB4
	Order	1	7.00	9.55	0.003	$1^{st}=4^{th}=2^{nd}<3^{rd}$	1	11.92	21.33	<0.001	$1^{st}=2^{nd}<4^{th}=3^{rd}$
	Residuals	110	80.58				112	62.56			
Invertebrate Biomass	Sub-Basin	3	1.29	0.48	0.697		3	4.10	1.62	0.189	
	Order	1	0.18	0.20	0.652		1	1.47	1.74	0.190	
	Residuals	110	98.53				112	94.43			
Predators	Sub-Basin	3	9.44	3.84	0.012	SB1=SB2=SB3<SB4	3	11.87	5.09	0.002	SB1=SB2 <SB3=SB4

	Order	1	0.45	0.55	0.461		1	1.12	1.45	0.232	
	Residuals	110	90.11				112	87.01			
Shredders	Sub-Basin	3	13.89	5.99	0.001	SB2=SB1=SB4<SB3	3	9.76	4.09	0.008	SB4=SB1 =SB2<SB3
	Order	1	1.12	1.45	0.232		1	1.26	1.58	0.211	
	Residuals	110	84.99				112	88.99			
Gatherer-Collectors	Sub-Basin	3	22.51	10.68	<0.001	SB4=SB3<SB2=SB1	3	27.15	14.29	<0.001	SB3=SB2 =SB4<SB1
	Order	1	0.23	0.33	0.656		1	1.91	3.02	0.085	
	Residuals	110	77.26				112	70.93			
Filtering-Collectors	Sub-Basin	3	11.14	4.89	0.003	SB1<SB3=SB4=SB2	3	10.61	4.75	0.004	SB1=SB2 =SB3<SB4
	Order	1	5.23	6.88	0.010	1st = 2nd = 4th <3rd	1	5.98	8.03	0.005	1st<4th=2nd=3rd
		110	83.63				112	83.41			
Scrapers	Sub-Basin	3	21.70	10.21	<0.001	SB3=SB4=SB2<SB1	3	11.79	5.07	0.002	SB3=SB4 =SB2<SB1
	Order	1	0.34	0.47	0.493		1	1.50	1.94	0.166	
	Residuals	110	77.96				112	86.71			
ATP	Sub-Basin	3	26.40	13.19	<0.001	SB2=SB1=SB4<SB3	3	17.59	8.07	<0.001	SB2=SB1 =SB3<SB4
	Order	1	0.20	0.31	0.581		1	1.06	1.46	0.230	
	Residuals	110	73.40				112	81.35			
Ergosterol	Sub-Basin	3	2.25	0.87	0.457		3	9.82	4.07	0.009	SB1=SB2 =SB3<SB4S
	Order	1	3.27	3.81	0.053		1	0.10	0.13	0.720	
	Residuals	110	94.47s				112	90.08			

Biotic Community

The density of invertebrates was higher in 4th order streams for both detritus types (means of 27 and 21 ind/g for *E. cloeziana* and *I. laurina*, respectively). Differences for *I. laurina* were observed among sub-basins, with high values in SB3 (mean 24 ind/g) and SB4 (mean 26 ind/g). The number of *taxa* was significantly different among the stream orders and sub-basins, with the highest values in SB3 (mean of 7 and 6 *taxa* for *E. cloeziana* and *I. laurina*, respectively) and SB4 (mean of 6 *taxa* for *E. cloeziana* and *I. laurina*), primarily in 3rd order streams for both detritus types (mean of 7 and 6 *taxa* for *E. cloeziana* and *I. laurina*, respectively). However, the biomass (total mean 0.003 and 0.002 ind/g for *E. cloeziana* and *I. laurina*, respectively) did not differ among the stream orders or sub-basins for either of the detritus types. The high variances in density, richness and biomass (in terms of the percentage of the sums of squares) were explained by differences in the sub-basins for both detritus types, except for the density of invertebrates in *I. laurina* (Figure S2 and S3, Table 2).

The functional trophic groups differed significantly among stream orders only for the filtering-collectors, with the highest values in the 3rd order streams (mean 20% for *E. cloeziana* and *I. laurina*) and the lowest in the 1st order streams (mean of 10% for *E. cloeziana* and *I. laurina*) for both detritus types. The relative abundance of predators was higher in SB3 (mean 24 and 25% for *E. cloeziana* and *I. laurina*, respectively) and SB4 (mean of 32 and 34% for *E. cloeziana* and *I. laurina*, respectively), whereas the values for shredders were higher in SB3 (mean of 27 and 25% for *E. cloeziana* and *I. laurina*, respectively) for both detritus types. However, SB1 exhibited high abundances of gathering-collectors (mean of 41 and 50% for *E. cloeziana* and *I. laurina*, respectively) and scrapers (mean of 18 and 16% for *E. cloeziana* and*I. laurina*, respectively) but a low abundance of filtering-collectors (mean of 6 and 8% for *E. cloeziana* and *I. laurina*, respectively) for both detritus types. The high variance in the relative abundance for all functional trophic groups (by the sums of squares) was also explained by changes in sub-basins for both detritus types (Figure S2 and S3, Table 2, Figure 3B and 4B).

The ATP values differed only among sub-basins for both detritus types, with the highest values in SB3 (mean of 2155.8 nmoles/g AFDM) for *E. cloeziana* (total mean 991.8 nmoles/g AFDM) and in SB3 (mean 633.3 nmoles/g AFDM) and SB4 (mean 2023.9 nmoles/g AFDM) for*I. laurina* (total mean 847.8 nmoles/g AFDM) (Table 2; Fig. 2C and 3C). There were no differences in the ergosterol concentrations among the hydrological stream orders and sub-basins for *E. cloeziana* (total mean 541.2 μg/g). However, we found higher ergosterol concentrations for *I. laurina* (total mean 382.9 μg/g) in SB3 (mean 392.6 μg/g) and SB4 (mean 464.1 μg/g), although they did not differ among the hydrological stream orders. The variances in ATP and ergosterol concentrations were also explained by changes in sub-basins for both detritus types, except for ergosterol in *E. cloeziana*, which showed a high level of variation with stream order (Figure S2 and S3, Table 2, Figure 3D and 4D).

DISCUSSION

Scale Analysis

The instantaneous discharge, water velocity, turbidity and nitrogen series (nitrate and nitrite) were more influenced by changes in stream order (high heterogeneity), with higher values downstream (increasing from 1st to 4th order), as expected according to Vannote et al. [1], except that nitrites exhibited the inverse pattern. The finding of relatively few influences at local scales can be explained by the large discontinuities inherent in smaller geomorphological units (habitat patches create discontinuities in space) that increase the potential influence from the local characteristics of the environment [4], [5], [14], [18]. Flow changes, for example, create hydrological discontinuities along stream corridors and isolate habitats. However, the other abiotic variables were influenced by changes in sub-basins (high homogeneity) that correspond to regional scales [19], [21]. These factors worked at the watershed level and may increase its fragility in the face of intense climatic changes because the climate is the primary controlling factor at large scales [22]. The basins are influenced by environmental factors that systematically change across longitudinal (upstream/downstream), vertical (sediment/water) and lateral (terrestrial/aquatic) gradients, forming different spatial and temporal patterns at regional and local scales [5], [17], [18].

As expected, the high-quality detritus (*E. cloeziana*) showed more rapid leaf breakdown rates (*k*) than the low-quality (*I. laurina*) detritus, indicating that the rate could be driven by micro-scale processes [14]. These findings represent important evidence that the riparian vegetation could be responsible for determining the ecosystem characteristics [30], [31], [59], as also proposed by Gonçalves et al. [35] for tropical systems, highlighting the need to study this vegetation. For example, if riparian vegetation is composed of plant species that have a higher stoichiometric ratio (higher quality), we expect more rapid response for organic matter cycling (higher decomposition). This direct relationship between terrestrial and aquatic ecosystems demonstrates that any modification in a riparian ecosystem would affect the function, primarily in areas composed of palatable plants (high quality and decomposition), corroborating the work of Frauendorf et al. [60]. Thus, lower quality vegetation (slower decomposition) will be less sensitive to other factors, and this may explain the resistance of the Brazilian savanna in comparison with other Brazilian tropical systems [61]. Moreover, the leaf breakdown at the macro-scale was more influenced by the regional scale (sub-basin) than by the local scale (stream order) [14], confirming our hypothesis. This result indicates that the patterns observed in studies covering a given time period (timely studies), common in tropical literature (for more see also [13], [32]), cannot be generalized from local to regional scales [17], [29], [62] or to whole watersheds [14]–[16], [18]. In addition, this finding may indicate fragility in the upstream basins due to the slower leaf breakdown rates. However, the upstream area is a source of nutrients and organisms for the downstream basins [1]. The upstream basins can give support productivity and may be responsible for extending the depuration capability of the system (as represented by the microbial pools) and maintaining functionality downstream [60]. Therefore, we believe that the association between detritus quality (important at the micro-scale) and the environmental features of the watershed (important at the macro-scale) is responsible for shaping organic matter cycling in the watershed and should be further investigated in future studies.

Previous studies indicating that the microbial community is the principal decomposer [6], [30],[35], are confirmed by our results, as we found that the high-quality leaf species was also more susceptible to leaching and microbial action, whereas the low-quality leaf species was influenced primarily by fungal colonization. Both leaf samples were consumed by shredders, but a higher abundance of shredders was observed in the *E. cloeziana* detritus. We also found a higher loss of mass due to the high water velocity (mechanical fragmentation and leaching), dissolved oxygen and temperatures, which accelerate biological metabolism [37], [63]. The detritus quality is important only for defining the local rates and their pathways for leaf breakdown [6], [35], [64]. However, the detritus quality has little influence on the general pattern along the "riverscape" and at any specific scale [15]. Therefore, based on an analysis of the samples after a certain percentage of mass has been lost (25, 75 and 50%) and not simply at predefined time points (e.g., 7, 15, 30 days), it is possible to show a clear colonization effect independent of quality. To be sure, detritus quality is a highly important determinant of the abundance of shredders and explains the importance of shredders for both detritus types [39]. We cannot study the variations associated with spatial scale in terms only of the local context because there are many factors in the ecological levels (community and ecosystem) that are responsible for variability found in the large scale [18], [29]. However, the local approach has been used in all previous tropical studies of leaf breakdown [12], [13].

Leaf Breakdown Rates

In agreement with the proposal of Gonçalves et al. [13] for tropical systems, the *k* values of *E. cloeziana* were classified as intermediate ($-0.0173 > k > -0.0041$), and those of *I. laurina* were classified as slow ($k < -0.0041$) for all sampling sites, indicating a strong influence of detritus quality. The high leaching due to the high solubility of polyphenols and tannins (secondary compounds) in *E. cloeziana* can accelerate the decay rate [64], [65]. Therefore, the rapid leaching of these secondary compounds, which has an inhibitory effect on detritivores, as well as the lower hardness of *E. cloeziana*, did not limit biotic colonization due to the low residence time in this type of detritus [35], [64]. Additionally, *Eucalyptus* sp. (an exotic species) is rapidly colonized and decomposed in the Brazilian savannah. It is possible that this pattern is due to the high quality of *Eucalyptus* relative to native species [35], [39], [66].

In contrast, we found lower breakdown rates for *I. laurina* which were most likely a consequence of a high content of structural compounds (lignin and cellulose) and relative hardness (cuticle thickness), hindering the release of other chemical compounds (e.g., polyphenols, nitrogen and phosphorus [35], [65], [66]). Therefore, the chemical characteristics of detritus determine the speed of processing (primarily at local scales), showing that leaf breakdown rates increase with quality and palatability [6], [7], [35], [39]. Detritus quality is of lower importance when if it is observed at different scales. In the study area, we observed that the regional scale is decisive for driving the general pattern of this important ecological process along the "riverscapes" [21], [22].

Abiotic and Biotic Variables in Leaf Breakdown

The natural environmental changes that occur across stream orders [4] were not sufficient to modify the remaining mass, and the local scale could not affect the decomposer communities (shredders and microorganisms) for either of the detritus types. Decomposer communities are the driving factors for leaf breakdown, and their absence leads to similar breakdown patterns along the stream order gradient [6]. This finding might indicate that ecological functioning in headwater streams (1st to 3rd order) was similar within the same sub-basin [1]. However, this process might change over a large spatial gradient, as represented by the sub-basin scale[22]. Nevertheless, increases in richness and the density of invertebrates and a decrease in the abundance of filtering-collectors for both detritus types were observed across this large spatial gradient. These results demonstrate that these variables had no effect on leaf breakdown [52], [53].

The values of the remaining mass for both detritus types were lower in SB4 and SB3 (high decomposition) than in the other sub-basins. It is probable that the reason for this difference was the higher temperatures, dissolved oxygen concentrations and water velocities resulting from the microclimate of the geographic location (within a valley). Therefore, the higher temperatures [63] and oxygen concentrations [37] observed in SB4 and SB3 may elevate the metabolic activity of the decomposer community [9], especially microorganisms. The higher metabolic activity of the decomposer community, associated with high water velocity (mechanical fragmentation and leaching), which increased the degree of physical abrasion [67], accelerated the leaf breakdown rates. In SB4 and SB3, higher density and richness of invertebrates and higher shredder abundance, with the greatest densities in *E. cloeziana*, were also observed. The shredders directly utilize leaf tissues for feeding, and increasing biological fragmentation [7], [42] can also accelerate the leaf breakdown rates [26]. Certain shredders in these locations (genus *Phylloicus*) can build their capsules from leaf tissue, and this use of leaf material also contributes to fragmentation [52], [53].

The relative abundance of shredders was influenced by variation, primarily among the sub-basins. A greater relative abundance of shredders in comparison with other tropical systems was observed in SB3 for both detritus types [10], [39], [42]. A low relative abundance of shredders was found in the other sub-basins. This result is consistent with the findings of previous studies in the Cerrado [13], [36], [39]. The importance of shredders for leaf breakdown is unclear in the tropics due to their low abundance or absence in these streams [10], [39],[42], but studies have shown little effect in tropical streams [11], [33]. From a global perspective, the strong effects observed in the current study were most likely due to the preference of shredders (primarily Trichoptera and Plecoptera) for high altitudes (due to the lower temperatures) in tropical regions [42]. The preference of this group for *E. cloeziana*indicates that detritus quality is also important [10] and that the composition of the vegetation influences the functioning of aquatic systems. Therefore, higher altitudes (low temperature[42]), high dissolved oxygen, the composition of the flora (ideally including *E. cloeziana* [10]) and moderate values of nutrient concentrations in the water [62] favor a high abundance of shredders. The predominance in the Cerrado of leaves that are low in nutrients [66], is associated with hydric and thermic stress and could be responsible for the absence or low abundance of shredders found in most tropical streams [10], [42].

The high-quality detritus (*E. cloeziana*) was shown to be most influenced by the total microbial community and the low-quality detritus (*I. laurina*) by the fungal community in SB4 and SB3 (high decomposition). *E. cloeziana* has elevated amounts of labile compounds, facilitating the activity of bacteria (rapid life cycles) that use compounds derived from the leaching of the leaves of labile detritus as their preferred resource [68]. These bacteria could be important during leaf breakdown and not only, as observed by several authors, at the early stages of the process [6], [13], [35]. However, this pattern is rarely observed in tropical streams. In lower-quality detritus (*I. laurina*), we observed an interaction between the biomass of the two microbial communities, highlighting fungi as the principal component [63]. Due to its high capacity to metabolize refractory molecules (e.g., cellulose and lignin) and to decompose them, the fungal community is the primary decomposer in tropical streams, and this principle explains the great significance of fungi in *I. laurina* decomposition [6], [35], [68]. Fungal action can increase the palatability of detritus, as well as its nutritional quality, for other decomposers, and the high biomass of fungi might be another factor responsible for the higher abundance of shredders in these sub-basins [7], [35].

In general, we conclude that variations in scale contribute to the variation in the leaf breakdown rate, highlighting the importance of similar studies of this type that determine effects at different scales. The variability of the physical structure of streams (primarily temperature, dissolved oxygen and nutrients) accelerates leaf breakdown from upstream to downstream, but this process was only demonstrated at the sub-basin scale in the location studied, partially corroborating the initial hypothesis. The replacement of shredder invertebrates by microorganisms was observed but was contrary to the prediction of our hypothesis. Shredders were favored by microorganisms (primarily in *E. cloeziana*), with stronger interactions between them than those previously found to drive leaf breakdown rates. Based on our interest in the influence of spatial structure on ecosystem functions, we observed that watersheds with warmer microclimates and streams with higher nutrient levels and oxygen in the water could be accelerating the metabolism of the ecosystem in the watershed, with increased negative effects downstream. For the management of tropical watersheds, we noted that the upstream areas are more fragile and sensitive to environmental impacts but show greater importance in the cycling of nutrients. We performed analyses within and between spatial scales to assess the relative importance of various watershed scales in determining the local breakdown rate for leaves. Local characteristics are responsible for the diversification of this process across the "riverscape", and high heterogeneity underscores the difficulty of making predictions based on local studies.

ACKNOWLEDGMENTS

The first author also received a scholarship from CAPES during his PhD research in the Graduate Program in Ecology of UFSC. We are also grateful to Cássia Alves Lima for valuable logistical support for the field and laboratory work.

REFERENCES

1. Vannote RL, Minshall GW, Cummins KW, Sedell JR, Cushing CE (1980) River Continuuum Concept. Canadian Journal of Fisheries and Aquatic Sciences 37: 130–137. doi: 10.1139/f80-017
2. Fabre E, Chauvet E (1998) Leaf breakdown along an altitudinal stream gradient. Archiv für Hydrobiologie 141: 167–179.
3. Statzner B, Higler B (1985) Questions and comments on the river continuum concept. Canadian Journal of Fisheries and Aquatic Sciences 42: 1038–1044. doi: 10.1139/f85-129
4. Benda L, Poff NL, Miller D, Dunne T, Reeves G, et al. (2004) The network dynamics hypothesis: How channel networks structure riverine habitats. BioScience 54: 413–427. doi: 10.1641/0006-3568(2004)054[0413:tndhhc]2.0.co;2
5. Larned ST, Datry T, Arscott DB, Tockner K (2010) Emerging concepts in temporary-river ecology. Freshwater Biology 55: 717–738. doi: 10.1111/j.1365-2427.2009.02322.x
6. Gessner MO, Chauvet E, Dobson M (1999) A perspective on leaf litter reakdown in streams. Oikos 85: 377–384. doi: 10.2307/3546505
7. Graça MAS, Cressa C, Gessner MO, Feio MJ, Callies KA, et al. (2001) Food quality, feeding preferences, survival and growth of shredders from temperate and tropical streams. Freshwater Biology 46: 947–957. doi: 10.1046/j.1365-2427.2001.00729.x
8. Mathuriau C, Chauvet E (2002) Breakdown of leaf litter in a neotropical stream. Journal of the North American Benthological Society 21: 384–396. doi: 10.2307/1468477

9. Suberkropp K, Chauvet E (1995) Regulation of Leaf Breakdown by Fungi in Streams: Influences of Water Chemistry. Ecology 76: 1433–1445. doi: 10.2307/1938146
10. Wantzen KM, Wagner R (2006) Detritus processing by invertebrate shredders: a neotropical-temperate comparison. Journal of the North American Benthological Society 25: 216–232. doi: 10.1899/0887-3593(2006)25[216:dpbisa]2.0.co;2
11. Moulton TP, Magalhaes-Fraga SAP, Brito EF, Barbosa FA (2010) Macroconsumers are more important than specialist macroinvertebrate shredders in leaf processing in urban forest streams of Rio de Janeiro, Brazil. Hydrobiologia 638: 55–66. doi: 10.1007/s10750-009-0009-1
12. Abelho M (2001) From litterfall to breakdown in streams: a review. Scientific World Journal 1: 656–680. doi: 10.1100/tsw.2001.103
13. Gonçalves JFJ, Martins RT, Ottoni BMP and Couceiro SRM (2013) Uma visão sobre a decomposição foliar em sistemas aquáticos brasileiros. In: N Hamada, J. L Nessimian and R. B Querino, editors. Insetos aquáticos: biologia, ecologia e taxonomia.
14. 14.Wiens JA (1989) Spatial Scaling in Ecology. Functional Ecology 3: 385–397. doi: 10.2307/2389612
15. Tiegs SD, Akinwole PO, Gessner MO (2009) Litter decomposition across multiple spatial scales in stream networks. Oecologia 161: 343–351. doi: 10.1007/s00442-009-1386-x
16. Lowe WH, Likens GE, Power ME (2006) Linking scales in stream ecology. Bioscience 56: 591–597. doi: 10.1641/0006-3568(2006)56[591:lsise]2.0.co;2
17. Schneider DC (2001) The rise of the concept of scale in ecology. BioScience 51: 545–554. doi: 10.1641/0006-3568(2001)051[0545:trotco]2.0.co;2
18. Wiens JA (2002) Riverine landscapes: taking landscape ecology into the water. Freshwater Biology 47: 501–515. doi: 10.1046/j.1365-2427.2002.00887.x
19. Allen TFH, Starr TB (1982) Hierarchy: Perspectives for Ecological Complexity. Chicago: University of Chicago Press.
20. Del Arco AI, Ferreira V, Graça MAS (2012) The performance of biological indicators in assessing the ecological state of streams

with varying catchment urbanisation levels in Coimbra, Portugal. Limnetica 31: 141–154.

21. Frissell CA, Liss WJ, Warren CE, Hurley MD (1986) A Hierarchical Framework for Stream Habitat Classification: Viewing Streams in a Watershed Context. Environmental Management 10: 199–214. doi: 10.1007/bf01867358
22. Allan JD (2004) Landscapes and Riverscapes: The Influence of Land Use on Stream Ecosystems. Annual Reviews of Ecology, Evolution, and Systematics 35: 257–284. doi: 10.1146/annurev.ecolsys.35.120202.110122
23. Royer TV, Minshall GW (2003) Controls on leaf processing in streams from spatial-scaling and hierarchical perspectives. Journal of North American Benthological Society 22: 352–358. doi: 10.2307/1468266
24. Irons JG, Oswood MW, Stout RJ, Pringle CM (1994) Latitudinal patterns in leaf litter breakdown: is temperature really important? Freshwater Biology 32: 401–411. doi: 10.1111/j.1365-2427.1994.tb01135.x
25. Minshall GW, Petersen RC, Cummins KW, Bott TL, Sedell JR, et al. (1983) Interbiome Comparison of Stream Ecosystem Dynamics. Ecological Monographs 53: 2–25. doi: 10.2307/1942585
26. Sponseller RA, Benfield EF (2001) Influences of Land Use on Leaf Breakdown in Southern Appalachian Headwater Streams: A Multiple-Scale Analysis. Journal of the North American Benthological Society 20: 44–59. doi: 10.2307/1468187
27. Hagen EM, Webster JR, Benfield EF (2006) Are leaf breakdown rates a useful measure of stream integrity along an agricultural landuse gradient? Journal of the North American Benthological Society 25: 330–343. doi: 10.1899/0887-3593(2006)25[330:albrau]2.0.co;2
28. Hieber M, Gessner MO (2002) Contribution of Stream Detrivores, Fungi, and Bacteria to Leaf Breakdown Based on Biomass Estimates. Ecology 83: 1026–1038. doi: 10.2307/3071911
29. Young RG, Collier KJ (2009) Contrasting responses to catchment modification among a range of functional and structural indicators of river ecosystem health. Freshwater Biology 54: 2155–2170. doi: 10.1111/j.1365-2427.2009.02239.x

30. Cummins KW (1974) The importance of different energy sources in freshwater ecosystems. In: K. W Cummins, editor editors. Productivity of World Ecosystems. Washington: National Academy of Sciences. pp. 166.
31. Petersen RC, Cummins KW (1974) Leaf processing in a woodland stream. Freshwater Biology 4: 343–368. doi: 10.1111/j.1365-2427.1974.tb00103.x
32. Bianchini Jr I (1999) Aspectos do processo de decomposição nos ecossistemas aquáticos continentais. In: M. L. M Pompêo, editor editors. Perspectivas da Limnologia no Brasil. São Luis: Gráfica e Editora União. pp. 21–43.
33. Moretti M, Gonçalves JJF, Callisto M (2007) Leaf breakdown in two tropical streams: Differences between single and mixed species packs. Limnologica 37: 250–258. doi: 10.1016/j.limno.2007.01.003
34. Bruder A, Schindler MH, Moretti MS, Gessner MO (2013) Litter decomposition in a temperate and a tropical stream: the effects of species mixing, litter quality and shredders. Freshwater Biology: n/a-n/a.
35. Gonçalves JF Jr, Rezende RS, Martins NM, Gregorio RS (2012) Leaf breakdown in an Atlantic Rain Forest stream. Austral Ecology 37: 807–815. doi: 10.1111/j.1442-9993.2011.02341.x
36. Gonçalves JF Jr, Graça MAS, Callisto M (2007) Litter decomposition in a Cerrado savannah stream is retarded by leaf toughness, low dissolved nutrients and a low density of shredders. Freshwater Biology 52: 1440–1451. doi: 10.1111/j.1365-2427.2007.01769.x
37. Medeiros AO, Pascoal C, Graça MAS (2009) Diversity and activity of aquatic fungi under low oxygen conditions. Freshwater Biology 54: 142–149. doi: 10.1111/j.1365-2427.2008.02101.x
38. Wright MS, Covich AP (2005) Relative importance of bacteria and fungi in a tropical headwater stream: Leaf decomposition and invertebrate feeding preference. Microbial Ecology 49: 536–546. doi: 10.1007/s00248-004-0052-4
39. Gonçalves JF Jr, Rezende RS, França J, Callisto M (2012) Invertebrate colonisation during leaf processing of native, exotic and artificial detritus in a tropical stream. Marine and Freshwater Research 63: 428–439. doi: 10.1071/mf11172

40. Li AOY, Ng LCY, Dudgeon D (2009) Effects of leaf toughness and nitrogen content on litter breakdown and macroinvertebrates in a tropical stream. Aquatic Sciences 71: 80–93. doi: 10.1007/s00027-008-8117-y
41. Ligeiro R, Moretti MS, Goncalves JF Jr, Callisto M (2010) What is more important for invertebrate colonization in a stream with low-quality litter inputs: exposure time or leaf species? Hydrobiologia 654: 125–136. doi: 10.1007/s10750-010-0375-8
42. Boyero L, Pearson RG, Dudgeon D, Ferreira V, Graça MAS, et al. (2012) Global patterns of stream detritivore distribution: implications for biodiversity loss in changing climates. Global Ecology and Biogeography 21: 134–141. doi: 10.1111/j.1466-8238.2011.00673.x
43. Larned ST (2000) Dynamics of coarse riparian detritus in a Hawaiian stream ecosystem: a comparison of drought and post-drought conditions. Journal of the North American Benthological Society 19: 215–234. doi: 10.2307/1468066
44. Rueda-Delgado G, Wantzen KM, Tolosa MB (2006) Leaf-Litter Decomposition in an Amazonian Floodplain Stream: Effects of Seasonal Hydrological Changes. Journal of the North American Benthological Society 25: 233–249. doi: 10.1899/0887-3593(2006)25[233:ldiaaf]2.0.co;2
45. Sutherland WJ, Freckleton RP, Godfray HCJ, Beissinger SR, Benton T, et al. (2013) Identification of 100 fundamental ecological questions. Journal of Ecology 101: 58–67. doi: 10.1111/1365-2745.12025
46. Graça MAS, Barlocher F, Gessner MO (2005) Methods to Study Litter Decomposition. Dordrecht: Springer.
47. 47.Strahler AN (1952) Hypsometric (Area Altitude) Analysis of Erosional Topology. Geological Society of America Bulletin 63: 1117–1142. doi: 10.1130/0016-7606(1952)63[1117:haaoet]2.0.co;2
48. Golterman HL, Clymo RS, Ohnsta DMAM (1978) Methods for chemical analysis of freshwater. Oxford: Blackwell Scientific Publications.
49. 49.Koroleff F (1976) Determination of nutrients. In: K Grasshoff, editor editors. Methods of sea water analysis. Verlag Chemie Weinhein. pp. 117–181.

50. Strickland JDH, Parsons TR (1960) A manual of seawater analysis. Ottawa: Fisheries Research Board of Canada. 188 p.
51. 51.Pérez GP (1988) Guía para el studio de los macroinvertebrados acuáticos del departamento de Antioquia. Bogotá: Editorial Presencia Ltda.
52. 52.Merrit RW, Cummins KW (1996) An introduction to the aquatic insects of North America. Dubuque: Kendall/Hunt Publishing Company.
53. 53.Cummins K, Merritt R, Andrade P (2005) The use of invertebrate functional groups to characterize ecosystem attributes in selected streams and rivers in south Brazil. Studies on Neotropical Fauna and Environment 40: 69–89. doi: 10.1080/01650520400025720
54. Bärlocher F, Graça MAS (2005) Total Phenolics. In: M. A. S Graça, F Barlocher and M. O Gessner, editors. Methods to Study Litter Decomposition. Dordrecht: Springer. pp. 97–100.
55. Gessner MO (2005) Proximate Lignin and Cellulose. In: M. A. S Graça, F Barlocher and M. O Gessner, editors. Methods to Study Litter Decomposition. Dordrecht: Springer. pp. 115–120.
56. Abelho M (2005) Extraction and quantification of ATP as a measure of microbial biomass. In: M. A. S Graça, F Bärlocher and M. O Gessner, editors. Methods to Study Litter Decomposition. Springer. pp. 223–229.
57. Gessner MO (2005) Ergosterol as a measure of fungal biomass. In: M. A. S Graça, F Barlocher and M. O Gessner, editors. Methods to Study Litter Decomposition. Dordrecht: Springer. pp. 189–195.
58. Crawley MJ (2007) The R Book. England: John Wiley & Sons Ltd.
59. Naiman RJ, Décamps H (1997) The ecology of interfaces: riparian zones. Annu Rev Ecol Syst 28: 621–658. doi: 10.1146/annurev.ecolsys.28.1.621
60. Frauendorf TC, Colón-Gaud C, Whiles MR, Barnum TR, Lips KR, et al. (2013) Energy flow and the trophic basis of macroinvertebrate and amphibian production in a neotropical stream food web. Freshwater Biology 58: 1340–1352. doi: 10.1111/fwb.12131
61. Carvalho FMV, De Marco P, Ferreira LG (2009) The Cerrado into-pieces: Habitat fragmentation as a function of landscape use in the savannas of central Brazil. Biological Conservation 142: 1392–1403. doi: 10.1016/j.biocon.2009.01.031

62. Bae M-J, Kwon Y, Hwang S-J, Chon T-S, Yang H-J, et al. (2011) Relationships between three major stream assemblages and their environmental factors in multiple spatial scales. Annales de Limnologie - International Journal of Limnology 47: S91–S105. doi: 10.1051/limn/2011022
63. Gulis V, Suberkropp K (2003) Interactions between stream fungi and bacteria associated with decomposing leaf litter at different levels of nutrient availability. Aquatic Microbial Ecology 30: 149–157. doi: 10.3354/ame030149
64. Ardon M, Pringle CM (2008) Do secondary compounds inhibit microbial- and insect-mediated leaf breakdown in a tropical rainforest stream, Costa Rica? Oecologia 155: 311–323. doi: 10.1007/s00442-007-0913-x
65. Ardon M, Stallcup LA, Pringle CM (2006) Does leaf quality mediate the stimulation of leaf breakdown by phosphorus in Neotropical streams? Freshwater Biology 51: 618–633. doi: 10.1111/j.1365-2427.2006.01515.x
66. Oliveira-Filho AT, Ratter JA (1994) A study of the origin of central Brazilian forests by the analysis of plant species distribution patterns. Journal of Botany 52: 141–194. doi: 10.1017/s0960428600000949
67. Santos Fonseca AL, Bianchini I, Pimenta CMM, Soares CBP, Mangiavacchi N (2012) The flow velocity as driving force for decomposition of leaves and twigs. Hydrobiologia 703: 59–67. doi: 10.1007/s10750-012-1342-3
68. Weyers HS, Suberkropp K (1996) Fungal and Bacterial Production during the Breakdown of Yellow Poplar Leaves in 2 Streams. Journal of the North American Benthological Society 15: 408–420. doi: 10.2307/1467795

Chapter 6

Mercury in the Soil of Two Contrasting Watersheds in the Eastern United States

Douglas A. Burns[1], Laurel G. Woodruff[2], Paul M. Bradley[3], William F. Cannon[4]

[1]United States Geological Survey, Troy, New York, United States of America,

[2]United States Geological Survey, Mounds View, Minnesota, United States of America,

ABSTRACT

Soil represents the largest store of mercury (Hg) in terrestrial ecosystems, and further study of the factors associated with soil Hg storage is needed to address concerns about the magnitude and persistence of global environmental Hg bioaccumulation. To address this need, we compared total Hg and methyl Hg concentrations and stores in the soil of different landscapes in two watersheds in different geographic settings with similar and relatively high methyl Hg concentrations in surface waters and biota, Fishing Brook, Adirondack Mountains, New York, and McTier Creek, Coastal Plain, South Carolina. Median total Hg concentrations and stores in organic and mineral soil samples were three-fold greater at Fishing Brook than at McTier Creek. Similarly, median methyl Hg concentrations were about two-fold greater in Fishing Brook soil than in McTier Creek soil, but this difference was significant only for mineral soil samples, and methyl Hg stores were not significantly different among these watersheds. In contrast, the methyl Hg/total Hg ratio was significantly greater at McTier Creek suggesting greater climate-driven methylation efficiency in the Coastal Plain soil than that of the Adirondack Mountains. The Adirondack soil had eight-fold greater soil organic matter than that of the Coastal Plain, consistent with greater total Hg stores in the northern soil, but soil organic matter – total Hg relations differed among the sites. A strong linear relation was evident at McTier Creek ($r^2 = 0.68$; $p<0.001$), but a linear relation at Fishing Brook was weak ($r^2 = 0.13$; $p<0.001$) and highly variable across the soil organic matter content range, suggesting excess Hg binding capacity in the Adirondack soil. These results suggest greater total Hg turnover time in Adirondack soil than that of the Coastal Plain, and that future declines in stream water Hg concentrations driven by declines in atmospheric Hg deposition will be more gradual and prolonged in the Adirondacks.

INTRODUCTION

Soil is the largest reservoir of mercury (Hg) storage in global terrestrial ecosystems, and transfers to and from soil are pivotal in Hg cycling among vegetation, the atmosphere, groundwater, surface water, and the oceans [1]. A recent study estimates the global soil pool size (year 2000) at 240,000 Mg Hg, and indicates that Hg storage has increased by about 20% since 1840 largely due to atmospheric Hg deposition primarily derived from anthropogenic emissions sources such as coal burning, cement manufacturing, and other industrial and mining activities [2]. Soil Hg concentrations vary widely from 10 ng/g to 1000 ng/g in rural and remote areas, and from 100 ng/g to >10,000 ng/g in urban, industrial, and mineralized/mined lands [3], [4]. This Hg is largely bound to soil organic matter (SOM), and Hg concentrations are typically strongly related to measures of SOM or soil organic carbon (SOC) [5]–[7]. Under oxidized conditions, Hg (II) is strongly bound to organic thiol groups, which are typically present in excess of available Hg in most soil [8], [9]. Under anoxic conditions, meta-cinnabar and other sulfide minerals may also be important Hg forms [10], though the presence and abundance of these mineral forms in soil without a metal sulfide ore genesis is not clear [11]. Additionally, Hg may be adsorbed either specifically or non-specifically to SOM and to mineral surfaces such as iron and aluminum oxy-hydroxides [12].

Given the strong affinity of Hg for SOM, soil Hg concentrations are typically greatest where SOM is concentrated, in shallow O- or A-horizons, and these concentrations typically decrease with depth [7]. The O-horizon commonly has Hg concentrations an order of magnitude greater than that of the mineral-dominated B-horizon, though Hg/SOM is typically greater in the B horizon [7], [13], [14]. Similarly, organic-rich soil such as peat typically has greater Hg concentrations than those of soil dominated by inorganic material [6]. Soil in areas with high geogenic Hg concentrations can show the reverse of the typical depth pattern, with greater Hg concentrations in the deep soil than near the surface [15]. Despite the greater Hg concentrations commonly measured in surface horizons, deeper mineral horizons generally store more Hg because of the much greater mass of mineral-dominated subsoil in most landscapes [4], [6], [16]. Total soil storage of Hg can vary from a few mg/m^2 to >50 mg/m^2, and these values vary, similarly to those of Hg concentrations, as a function of source loads of atmospheric Hg deposition, the amount of SOM in soil, and geogenic concentrations in parent material [3], [6]. Disturbance factors related to fire history, agriculture, forest harvesting, mining, and other human land uses also affect Hg concentrations and storage in soil [17], [18].

Several studies have described strong relations between soil Hg concentrations and SOM/SOC, consistent with the strong affinity of Hg for binding with thiol groups [5], [7], [13],[19]. The slopes and intercepts of these Hg–SOM relations vary greatly, however, as a function of regional location, horizon and/or depth, landscape type, and vegetation type [6],[16], a finding that is expected given the heterogeneity of SOM and the excess of thiol-like Hg binding sites generally available [9].

Some study results are consistent with variation of Hg/SOM as a function of the rate of atmospheric Hg deposition [5], [15], although variation in geogenic Hg may be a contributing factor as well [20].

Although measurements of total Hg (THg) in soil provide important data necessary to better understand the Hg cycle, data on methyl Hg (MeHg) concentrations in soil are more pivotal to understanding bioaccumulation in terrestrial and aquatic food webs. The ratio of MeHg/THg varies widely in soil from <1% to >20% [4], [6], [8]. Wetland soil generally has greater MeHg concentrations and MeHg/THg than soil in steeper upland parts of landscapes, and more complex patterns are often superimposed on this general gradient. One study in a boreal forest catchment in Sweden found greater MeHg concentrations and MeHg/THg in soil at stream bank and near-stream locations than 20 m from the stream [8]. Another more recent investigation showed the highest MeHg concentrations in soil pore water where upland soil borders peaty wetland soil suggesting that the distal riparian area may have greater soil MeHg concentrations than the near-stream riparian area [21]. The MeHg content of soil is important because of the potential for mobilization into soil solution followed by surface and/or subsurface runoff and transport to local surface waters where bioaccumulation in aquatic food webs may then occur.

Despite many published studies on Hg in soil [3]–[7], little is known about controls on the relation of Hg concentrations and stores in soil to those of adjacent surface waters. Elevated Hg levels are often measured in surface waters with extensive wetlands where high THg and MeHg concentrations and storage in soil have been demonstrated, such as lake watersheds in the US Upper Midwest [22] and the Adirondacks of New York [16].

In this study, we compare Hg concentrations and storage in the soil of two watersheds, located in "hot spot" regions where Hg in predator fish and/or aquatic-feeding birds consistently exceed thresholds for human health warnings and for adverse effects on wildlife [23], [24], the Adirondack Mountains of New York, and the Coastal Plain of South Carolina [25], [26]. Soil THg and MeHg concentrations and storage are compared across three different landscape types in each watershed, uplands, riparian areas, and wetlands, because hydrologic/transport dynamics among these landscapes are important in mobilization of Hg to local streams and subsequent bioaccumulation in the aquatic ecosystem [25]–[27]. Our objectives are to describe how Hg concentrations and pool sizes in soil differ between these two watersheds and among different landscape types, and to discuss the implications of any differences that can be readily identified.

STUDY SITES

Soil samples were collected in two geographic locations: (1) Fishing Brook watershed in the Adirondack Mountains of New York, and (2) McTier Creek watershed in the Coastal Plain of South Carolina. A brief description of these study sites follows with an emphasis on soil; however, the reader is referred to [28] for more detailed descriptions of these study areas.

Fishing Brook (FB) is a 65.6 km^2 watershed in the westernmost headwaters of the Hudson River basin. The watershed is a heterogeneous landscape characterized by forested uplands, riparian floodplains, and scattered open water bodies. The watershed is dominantly forested (86.3%) by Adirondack Northern Hardwood Forest [29]. Sugar maple (*Acer saccharum*), yellow birch (*Betula alleghaniensis*) and American beech (*Fagus grandifolia*) are the dominant tree species below 980 m elevation, and balsam fir (*Abies balsamea*) and red spruce (*Picea rubens*) are the dominant species above this elevation [29]. Additionally, balsam fir and red spruce are dominant in wet soil at the base of hillslopes and in riparian areas adjacent to streams and ponds. The watershed consists of 8.2% wetland area (National Wetlands Inventory (NWI) approach, [30]), and scrub-shrub is the dominant wetland type, with speckled alder (*Alnus incana*) a dominant species [31]. Open water is 2.7% of watershed area, with numerous small ponds and lakes. Only 0.7% of the watershed is considered developed, as low density residential land use, some dirt roads, and one highway (Rte. 28N).

Upland soils in the FB watershed are mainly Spodosols developed in coarse, bouldery glacial till derived from the Precambrian metamorphic bedrock of the region [32]. The most common soil types found in the watershed include the Skerry, Becket, Adirondack, and Tunbridge Soil Series, generally well-drained, bouldery sandy loams classified as Haplorthods [33]. The most common soil types in riparian wetland areas are broadly classified as Entisols or Histosols and include the Wonsqueak, Rumney, and Bucksport Soil Series, mixtures of loamy sand or silt, muck, and peat with a water table that ranges from just below land surface during the growing season, to near or above the land surface during the dormant season [33]. These riparian soil types are generally developed in fluvial or glaciofluvial sediments, and thus are not as rich in organic matter as the deep peat riparian wetlands in the western Adirondacks described in [34].

McTier Creek (MC) is a 79.4 km^2 watershed (defined by the USGS stream gage near New Holland, SC) tributary to the Edisto River, and located in the Inner Coastal Plain. The watershed consists of 49.6% forested land cover, of which loblolly pine (*Pinus taeda*) is most common. About 6.4% of the watershed is wetland area (NWI approach, [35]), and bottomland hardwoods such as black tupelo (*Nyssa sylvatica*) are dominant in floodplain areas where these wetlands are generally found. Additionally, 1.2% of the watershed consists of open water, 14.9% is agricultural land, and 4.7% is developed [28]. The remaining land cover is mainly herbaceous upland.

Soil at MC is developed in sandy fluviomarine deposits and is generally classified as an Ultisol on ridges and hillslopes and as an Entisol or Inceptisol in riparian floodplains [36]. The Troup, Vaucluse, and Ailey Series are those most common on ridges and hillslopes, generally an excessively drained sandy soil with clay-rich kaolinitic deposits found at depth in isolated areas [36]. Floodplains are dominated by the poorly drained Bibb Series loamy sand and Johnston Series mucky loam, the latter of which sometimes has standing water at the surface signifying wetlands [36].

A comparison of the landscape and soil of these two watersheds reveals some similarities and some sharp differences. Both watersheds are dominated by upland forests, although the FB landscape is generally steeper (median slope = 12.5%) than that of MC (median slope = 5.4%), and MC also includes almost 20% developed land and agriculture, whereas FB has <1% developed land. Narrow floodplains dominated by sandy and loamy deposits as well as some mucky or peaty soil are prevalent in both watersheds, and riparian wetlands are commonly found in these floodplains. Riparian area defined as all land within an elevation of 0.65 m of the stream network [37] is 14.1% of the MC watershed area and 18.4% of the FB watershed area. When all wetland area located within this riparian delineation is excluded from these calculations, these riparian area values decrease to 9.3% of the MC watershed area and 12.3% of the FB watershed area. The two watersheds also have similar proportions of wetland area with a slightly greater value in the FB drainage (8.2%) than in the MC drainage (6.4%) according to NWI methods [35]. This difference narrows further (FB = 9.3%, MC = 8.2%) according to the wetland classification approach of the National Land Cover 2001 Dataset [38], but we apply NWI data in the current paper.

Perhaps the greatest difference between these two watersheds is their respective climates. FB experiences a cold, humid continental climate with mean annual air temperature of 4.9°C and mean annual precipitation of 1074 mm (Newcomb, NY, 1981–2010, [39]). Mean minimum daily air temperature is <0°C for >6 months of the year, much of the winter precipitation falls as snow, and spring snowmelt is a major feature of the annual hydrograph [40]. In contrast, MC experiences a subtropical moist climate with mean annual air temperature of 17.9°C and mean annual precipitation of 1276 mm, of which little falls as snow (Aiken, SC, 1981–2010, [41]). Stream runoff also differs greatly among these two regions. For example, the Edisto River near Givhans, SC of which MC is tributary has mean annual runoff of 273 mm (1981–2010, [42]), whereas the Hudson River at North Creek, NY of which FB is a tributary has mean annual runoff of 735 mm over the same period [42], a difference that is largely attributed to the role of snowmelt in New York, and to higher annual rates of evapotranspiration in South Carolina than in New York.

MATERIALS AND METHODS

Field

Soil sampling focused on the key landscape types in each watershed. At FB, samples were collected from upland hillslope areas in northern hardwood forest, along riparian areas in spruce-fir forest, and in wetlands [43]. A total of 163 soil samples were collected at 70 distinct locations during Sept. 2008 (Fig. 1). Sampling sites were on private land owned by Finch Pruyn Company (now Finch Company), and verbal permission was provided for collecting environmental samples including a key to access locked gates needed to reach many of the sites. At MC, samples were collected along two transects from upland hilltops and hillslopes, riparian floodplains, and wetlands [43]. A total of 81 samples were collected from 24 distinct locations in Nov. 2008 (Fig. 1). Sampling sites were on private land owned by the family of the late Senator Strom Thurmond who provided verbal permission for access and the collection of environmental samples. For the purposes of this analysis, soil data are distinguished for uplands, riparian floodplains, and wetlands in each watershed.

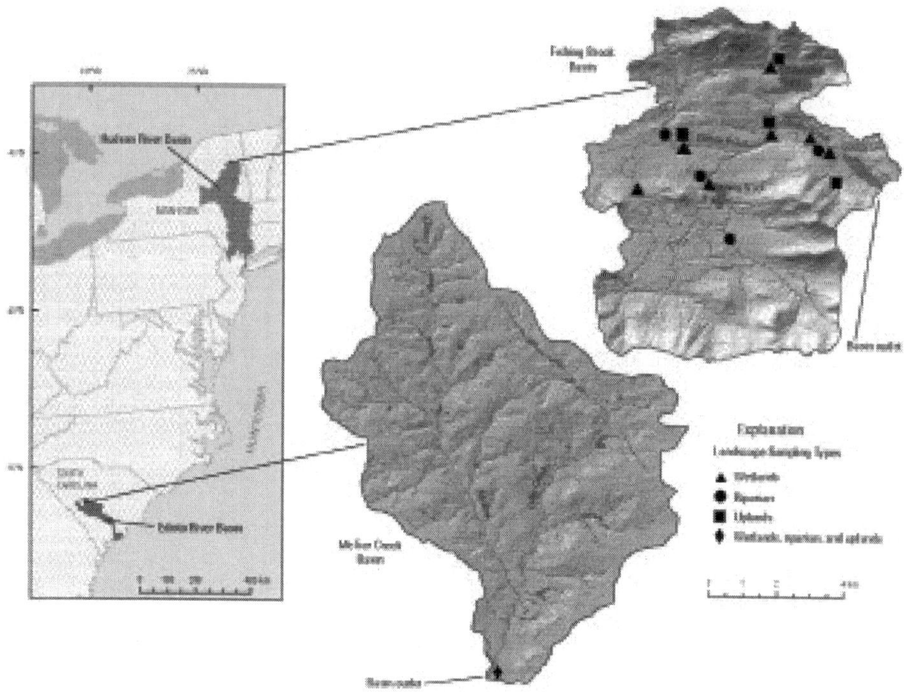

Figure 1. Study site maps. Locations of McTier Creek watershed within the Edisto River basin, and Fishing Brook within the Hudson River basin. Symbols indicate approximate locations where soil samples were collected in different landscape types as described in text. Note that only one symbol appears in the McTier Creek map because the 24 sampling sites were located too close together to be resolved in a map of the whole watershed.

Samples were collected with a shovel and hand auger and placed into sediment vials immediately upon collection. Nitrile gloves were used when handling soil to avoid contamination. At least two depths/horizons were sampled at each location, either an Oa (FB) or A (MC) horizon, and a B horizon. The B horizon represents a depth integrated sample that reflects the complete horizon to a depth of 40 cm; sampling did not attempt to make fine scale distinctions within the B horizon. At a few of the sites, samples were collected deeper in the profile, either the B/C or C horizon; E-horizon samples were also collected at both Fishing Brook and McTier. In wetland areas without well-developed soil horizons, samples were collected by depth interval. In such soil, the top 10 cm was typically equated to an Oa or A horizon for the purposes of this data analysis, and a second interval below this depth, typically 10 cm to about 20 to 40 cm, was equated to a B horizon for the purposes of this data analysis.

Laboratory

Oa horizon samples were air dried, milled, and split. A portion was set aside for Hg analysis, and the remainder was heated in an oven at 500°C for 13 hr to determine loss-on-ignition. For mineral soil horizons, samples were air dried and sieved to <2 mm. Total carbon (C) in mineral-dominated soil samples was determined by infrared detection of CO_2 in an automated C analyzer [44], carbonate C was determined by coulometric titration [45], and organic C was determined by difference. Organic C data were multiplied by 1.9 to convert these values to percent organic matter for the purposes of data analysis [46].

Mercury was analyzed by cold vapor atomic absorption spectrometry according to a method described in Smith et al. [47]. This method is a modification of U.S. Environmental Protection Agency Method 7471B [48], and is considered an analysis of total Hg. The method detection limit is 10 ng/g. A lab-homogenized sub-sample of soil that was frozen as soon as possible after collection was shipped to the laboratory on dry ice, and held frozen until analysis of MeHg by cold vapor atomic fluorescence spectrometry. Details of the MeHg analysis method, which has a 0.08 ng/g detection limit, are described in [49].

Laboratory quality assurance and quality control procedures for the analytical methods used in this study are described in [43] and [47], except for those of MeHg, which are described in[49]. Sixteen duplicate samples were collected in the FB watershed, and 7 were collected in the MC watershed. Precision for FB duplicates samples as represented by the mean standard deviation and mean relative standard deviation (in parentheses) of these duplicate analyses was 4.9 ng/g (8.4%) for total Hg, and 7.4% (22.0%) for percent organic matter. Precision for MC duplicates was 10 ng/g (21.1%) for total Hg, and 0.38% (12.4%) for percent organic matter. Duplicate analyses were not available for MeHg, but analytical precision data for multiple analyses of a reference solid sample indicate a relative standard deviation of 8.6% for the method used in this study [49].

Soil data used in this paper are available in [43], and can be freely downloaded.

Storage Estimates

THg and MeHg stores in soil were estimated at each study site. Estimates of soil element stores have high uncertainty due to a variety of factors including the presence of rocks, variations in the depth and thickness of soil horizons, high spatial heterogeneity of chemical and physical variables, and difficulties in accurately measuring bulk density, among others[50]. These sources of uncertainty should be carefully considered when comparing estimates of soil element stores among various studies in which assumptions can vary widely. In this investigation, we did not attempt to quantify soil rock content, a major source of uncertainty in element analyses [51], particularly in formerly glaciated terrain overlain by coarse till such as the FB watershed (rocks were not evident in the MC watershed soil). We developed estimates of THg and MeHg for the uppermost 40 cm of soil at each site, excluding fresh and partially decomposed litter (considered Oi and Oe horizons in some soil types), which was not sampled. These storage estimates were developed for the uplands, riparian area, and wetlands in each watershed using soil samples collected in each of these landscape types.

In Fishing Brook, the Oa horizon was assumed to be the uppermost 5 cm of soil in upland areas, the E horizon the next 5 cm, and the B horizon was assumed to include a depth interval of 10 to 40 cm, broadly consistent with observations in these Spodosols [33] and representing the mean of depth interval measurements made at the time of sampling. In the Entisol and Histosol soils that were dominant in riparian and wetland areas at FB and that frequently had poorly developed horizons, Hg data from Oa samples were applied to the depth interval of 0 to 10 cm, and B-horizon Hg data were applied to a depth interval of 10 to 40 cm.

In MC, the A horizon Hg data were applied to a depth interval of 0 to 4 cm in upland areas, 0 to 11 cm in riparian areas, and 0 to 20 cm in wetland areas based on the mean of depth measurements in each of these landscapes at the time of sampling. The B horizon Hg data were applied to the depth interval from the bottom of the A to 40 cm. The South Carolina soil survey typically includes an E horizon below the A in the soil types that dominate in uplands, and a C or more extensive A horizon below our A horizon classification in the soil types that dominate in riparian and wetland areas [36]. However, we recognize that some samples that were included in the B horizon category may actually be A, E, or C horizons samples based on this classification.

Bulk density was measured at 15 sites in A horizon soil at MC and at 38 sites in B horizon soil at FB. These measurements were made by sampling a known volume of soil and weighing this sample after drying. These measurements did not cover the full range of organic matter content measured in soil at these sites. In the case of MC, low organic matter soil (<5%) was not represented, whereas at FB, high organic matter soil (>24%) was not represented. Available bulk density measurements were averaged within 5 categories each representing a 5% range of soil organic matter (SOM) from 0% to 25% at FB, and at MC were averaged and placed into 4 categories each representing a 10% range up to 40% SOM (first category was 5% to 10%). These categorical data were best fit by a highly significant ($p<0.001$) power law regression of the form: bulk density = a * percent organic matter (–exp b). These equations were used to calculate bulk density values in the range of SOM >5% at MC and SOM<24% at FB. This approach provided good estimates of bulk density as demonstrated by mean deviations from measured values of +0.016 g/cm^3 at FB and –0.004 g/cm^3 at MC with no evident bias (organic matter vs. residuals linear regression, slope not significantly different than 0, $p>0.10$) across the OM range. For the SOM ranges of 0–5% at McTier Creek and 24–100% at Fishing Brook, a pedo-transfer equation was applied [52]. This approach is based on a simple mixing model between the bulk density values of 100% organic matter and that of 100% mineral soil and has proven useful in many previous investigations [53], [54]. The value used in this equation for 100% SOM at MC was 0.10 g/cm^3 derived from the power law regression, and the value for 100% mineral soil was 1.70 g/cm^3 based on sandy Coastal Plain B-horizon soil with little organic matter [55], [56]. The value used in this equation for 100% SOM at FB was 0.11 g/cm^3 based on data from Spodosols in the northeastern US [57] and the value used for 100% mineral soil was 1.99 based on the power law regression equation as it approaches 0% SOM (at 1% SOM).

Statistical Analyses

Seven of 163 soil samples collected at Fishing Brook had THg concentrations less than the 0.1 ng/g laboratory detection limit, and 6 of 81 soil samples collected at McTier Creek also had THg concentrations less than this detection limit. Therefore, non-parametric statistical analyses appropriate for censored data were used for any calculations that included THg data. Although there were no censored MeHg values, non-parametric statistical analyses were applied to these data as well because data were not normally distributed. The Mann-Whitney Rank-Sum test [58] was applied for pairwise comparisons, and the Kruskal-Wallis test for multiple comparisons [59]. Maximum likelihood regression was used instead of least squares linear regression, and a likelihood r^2 value calculated for each model [60]. Finally, relations among variables were determined by Spearman rank correlation analysis. All statistical analyses were performed with SigmaStat, part of the SigmaPlot software (v. 12, use of product names in this paper is for identification purposes only and does not imply endorsement by the U.S. Government), except for maximum likelihood regression, which was performed with SAS.

RESULTS

THg and MeHg concentrations and the THg/MeHg ratio for the two watersheds are compared first in Oa/A horizon samples and second in B horizon samples. These soil concentrations and ratios are also compared across landscape type within each watershed. The relations of Hg species concentrations to percent SOM are then presented. Finally, estimated storage of THg and MeHg in soil is presented and compared among the two watersheds.

Comparison of Soil Hg among Watersheds – Oa and A Horizon Soil

THg concentrations were significantly greater ($p<0.05$) in Oa/A horizon soil samples from FB than those from MC (Fig. 2A, Table 1). FB soil also had significantly greater THg concentrations than that of MC across all three landscape types. The median THg concentration of 200 ng/g at FB was more than three-fold greater than the median value of 60 ng/g at MC, and an approximate two-fold to four-fold difference was measured across the three landscape types. The differences in THg among soil from these two watersheds were so great, that the highest concentrations measured at MC were less than the median concentrations at FB for all samples as well as for those collected in each landscape type (Fig. 2A).

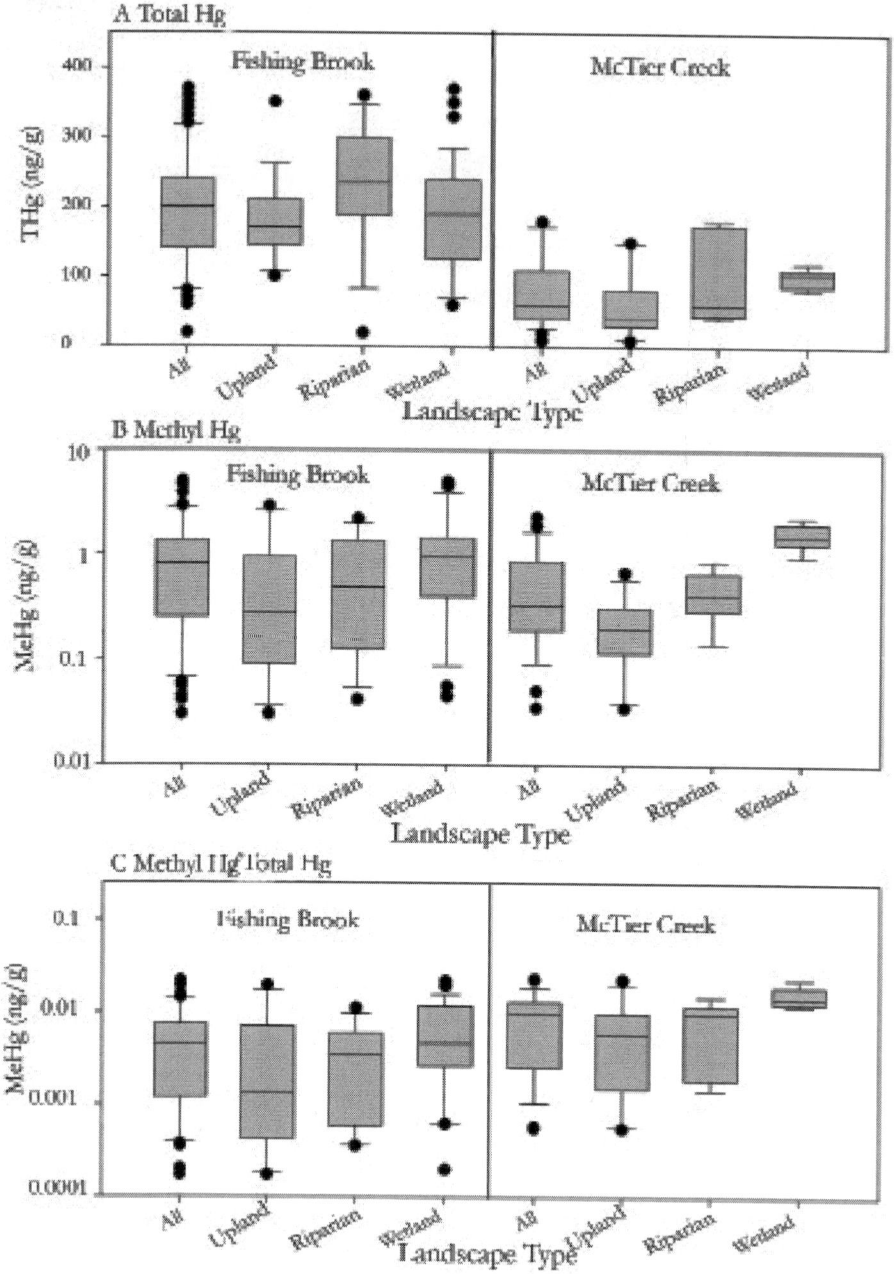

Figure 2. Mercury values in soil Oa/A horizon samples. Box plots of soil Hg concentrations measured in the Fishing Brook, NY and McTier Creek, SC watersheds for samples collected from Oa/A horizons. A Total Hg, B Methyl Hg, and C Methyl Hg/Total Hg.

Table 1. Median values of THg, MeHg, and MeHg/THg in Oa/A horizon soil samples from Fishing Brook, NY and McTier Creek, SC

Constituent	**Fishing Brook**				Maier Creek			
	All	Upland	Riparian	Wetland	All	Upland	Riparian	**Wetland**
THg(ng/g)								
n	71	17	16	38	27	12		6
median	200	170	235	190	60	40	60	105
MeHg								
n	52	11	12	29	27	12	9	6
median	0.81	0.28	0.48	0.94	0.33	0.20	0.42	1.50
MeHg/THg (%)								
n	52	11	12	29	27	12	9	6
median	0.44	0.13	0.35	0.46	0.98	0.57	0.98	1.40

The median MeHg concentration of 0.81 ng/g for all FB soil samples was more than two-fold greater than the median value of 0.33 ng/g for all MC samples, but the difference between these watersheds was not significant ($p = 0.133$; Table 1). The median value for upland and riparian samples was also slightly greater at FB than MC, but these differences were similarly not significant. In contrast, the median MeHg concentration of wetland soil samples collected at MC of 1.5 ng/g was greater than the value of 0.94 ng/g in FB wetland samples, a difference of marginal statistical significance ($p = 0.088$). There were fewer significant site differences for MeHg than for THg because of greater variation in the MeHg data (Fig. 2b); relative standard deviation values were 40.0% and 63.8% for THg at FB and MC, respectively, and increased to 109.4% and 99.5% for MeHg at these watersheds.

The MeHg/THg ratio showed the opposite pattern of the individual constituents with values two- to four-fold greater at MC than at FB (Table 1). This ratio was significantly greater for all samples and for those from the wetlands, but differences were not significant for riparian ($p = 0.110$) and upland ($p = 0.148$) soil. Though MeHg soil concentrations in the Coastal Plain watershed were generally less than those in the Adirondack watershed, MeHg/THg values were generally greater in the south.

Comparison of Soil Hg between Watersheds – B Horizon Soil

Soil Hg concentrations were generally less in B horizon soil than those in Oa/A horizon soil at both watersheds (Table 2, Figs. 2 and 3). The inter-watershed differences in Hg concentrations in the deeper B horizon soil were broadly consistent with the patterns in the shallower Oa/A horizon soil with some minor differences. THg concentrations were significantly greater in FB B horizon soil than those of MC for all samples and for those from uplands and riparian areas, but not for wetland samples. Median MeHg concentrations in B horizon samples were generally greater at FB than at MC as in Oa/A horizon samples. Whereas these inter-watershed differences were not statistically significant in Oa/A horizon soil, the differences were highly significant for all B horizon samples ($p<0.001$) and significant for wetland samples ($p = 0.023$), but not significant for those from uplands and riparian floodplains. Finally, MeHg/THg was also generally greater in B horizon soil at MC than at FB consistent with the findings for the shallow soil horizons. These differences were highly significant for upland soil ($p = 0.001$), of marginal significance for all soil samples ($p = 0.052$), but not significant for riparian and wetland samples.

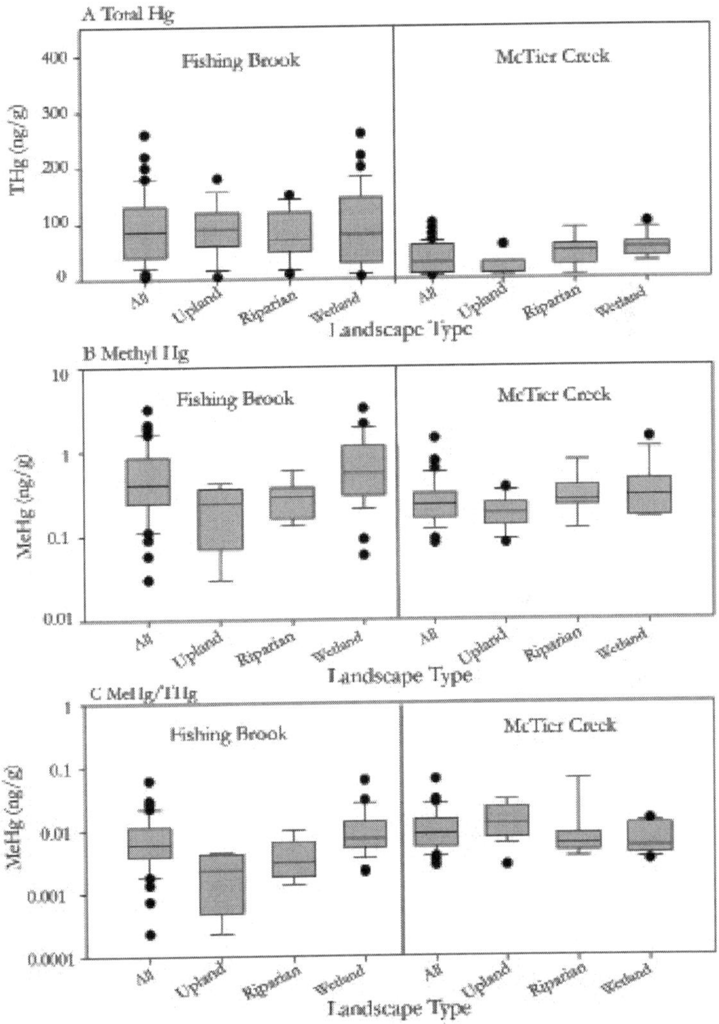

Figure 3. Mercury values in soil B horizon samples. Box plots of soil Hg concentrations measured in the Fishing Brook, NY and McTier Creek, SC watersheds for samples collected from B horizons. A Total Hg, B Methyl Hg, and C Methyl Hg/Total Hg.

Table 2. Median values of THg, MeHg, and MeHg/THg in B horizon soil samples from Fishing Brook, NY and McTier Creek, SC

Constituent	Fishing Brook				Maier Creek			
	All	**Upland**	**Riparian**	**Wetland**	**All**	**Upland**	**Riparian**	**Wetland**
THg(ng/g)								
n	70	17	16	37	37	19	9	14
median	85	90	70	80	80	10	50	55
MeHg								
n	40	5	7	28	27	19	9	14
median	0.40	0.24	0.29	0.54	0.54	0.18	0.26	0.28
MeHg/THg (%)								
n	52	5	7	7	28	19	9	14
median	0.44	0.23	0.31	0.31	0.71	1.20	0.60	0.53

Comparison of Soil Hg across Landscape Types

Within the FB watershed, THg concentrations in soil were similar across all landscape types for both Oa/A and B horizon samples (Figs. 2 and 3). MeHg and MeHg/THg values were also similar and statistically indistinguishable across all landscape types for Oa/A horizon soil samples. However, in B horizon samples, MeHg concentrations and MeHg/THg were significantly different across the three landscape types. Pairwise comparisons among the landscape types showed that wetlands>riparian areas = uplands, a consistent finding for both MeHg concentrations and MeHg/THg.

MC showed more differences in soil Hg concentrations as a function of landscape type than did FB (Figs. 2 and 3). All Hg measures were significantly different among the three landscape types in both Oa/A and B horizon samples. In pairwise comparisons, wetland values were nearly always significantly greater than those of the uplands, and the riparian and upland landscape types were most often statistically indistinguishable, despite median THg and MeHg values that were greater in riparian samples. The exception to these generalizations was MeHg/THg in the B horizon samples for which the upland ratio was significantly greater than the wetland ratio, whereas the other pairwise comparisons showed no significant differences.

Relations of Hg to Percent Soil Organic Matter

Some strong differences are evident in the relations of THg to percent SOM among these two watersheds and as a function of soil depth and landscape type (Fig. 4; Table 3). A highly significant ($p<0.001$) positive maximum likelihood regression relation was evident between THg and SOM for all soil samples collected at FB, but this relation explains only 13% of the variation in these data. At MC, an equally highly significant relation ($p<0.001$) explains 68% of the variation in these data (Fig. 4). Despite the significance of the regression relation at Fishing Brook, a LOESS smooth better fits the data pattern and confirms a generally non-linear THg-SOM relation that is highly variable throughout the range of SOM data; a generally linear increase in THg to an SOM value of about 15% ($r^2 = 0.46$, $p<0.001$), little change and much scatter in the range of 15% to about 60%, and a linear decrease in THg at SOM values >60% (Fig. 4A). In contrast, THg values at MC show much scatter at SOM values <20%, but generally increase with SOM at values >20%, though there are few samples with high SOM values (Fig. 4B). These data highlight the generally greater SOM in FB soil where values >90% were measured, whereas the greatest SOM values in soil samples at MC were about 40%.

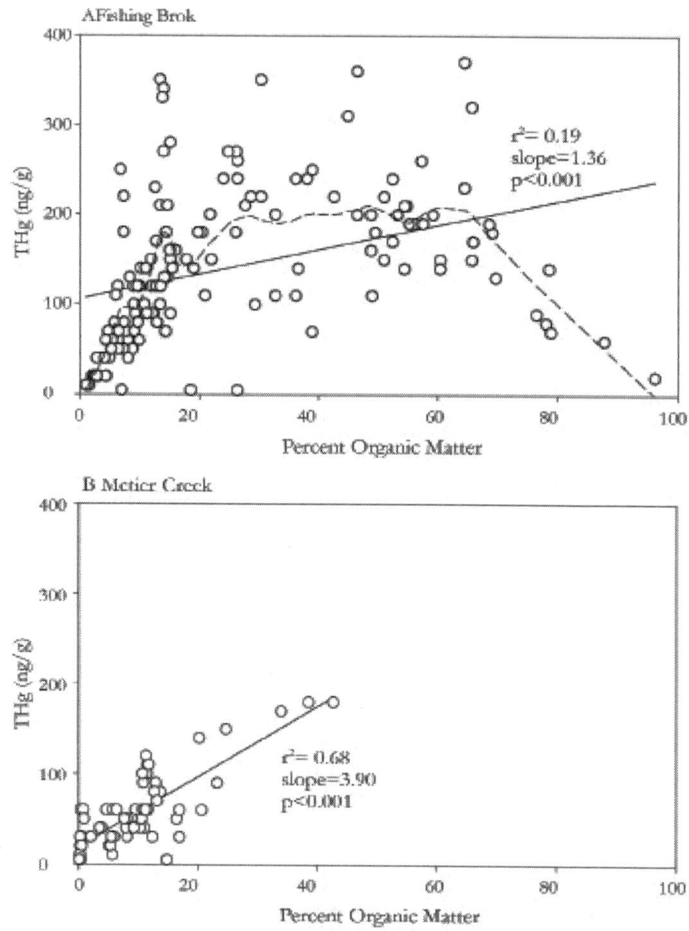

Figure 4. Total Hg and percent organic matter for all soil samples collected in each watershed. Solid line in each panel indicates a significant linear regression relation with r^2, slope, and p value. A Fishing Brook. Dashed line is a Loess smooth obtained by a first order polynomial fitted to a 20% moving window of the data. B McTier Creek.

Table 3. Spearman rank correlations of THg and percent organic matter and MeHg and percent organic matter in Oa/A horizon and B horizon soil at Fishing Brook, NY and McTier Creek, SC

Site/landscape Position	Oa/A Horizon				B Horizon			
	THg		MeHg		THg		Meng	
	n	r_s	n	r,	n	r_s		
Fishing Brook								
All	71	-0.315	52	-0.117	70	0.759	40	0.413
Upland	17	-0.129	11	0.036	17	**0.780**	5	-0.600
Riparian	16	0.093	12	0.14	16	0.793	7	-0.071
Wetland	38	-0.402	29	-0.205	37	**0.734**	28	0.516
McTier Creek								
All	27	0.654	27	-0.125	42	0.750	42	0.584
Upland	12	0.864	12	-0.070	19	0.694	19	0.312
Riparian	9	0.906	9	-0.460	9	0.252	9	0.429
Wetland	6	-0.232	6	-0.377	14	0.648	14	0.460

The inverse relation at FB of THg to SOM at high SOM values (as per LOESS smooth) results largely from the Oa/A horizon samples, and is only evident ($p<0.05$) among the wetland soil samples (Table 3). In contrast, a strong positive relation of THg and SOM is evident in the B horizon soil samples at Fishing Brook, and is consistent throughout all three landscape types. At MC, THg and SOM were significantly positively related in all landscape types except for the Oa/A horizon wetland samples, and in all landscape types except for the riparian in B horizon samples (Table 3). Overall, far fewer significant MeHg-SOM relations were evident in these data than were evident in the THg-SOM data. For all soil samples collected at each site, the relation between MeHg and SOM was not significant (Fig. 5). Among soil depths, there were no significant MeHg-SOM relations for any landscape type at either watershed (Table 3). In B horizon soil samples by contrast, MeHg was positively related to SOM among all samples for both watersheds. However, only the wetland soil at FB showed a significant MeHg-SOM relation, whereas the relation at MC was of marginal ($p = 0.094$) significance.

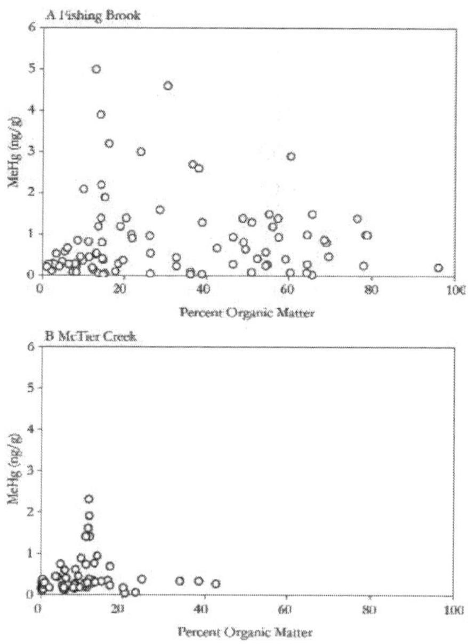

Figure 5. Methyl Hg and percent organic matter for all soil samples collected in each watershed. Absence of a line in panel indicates that a significant linear regression relation was not evident. A Fishing Brook, B McTier Creek.

Hg Storage in Soil of the Two Watersheds

Stores of THg in FB were two- to four-fold greater ($p<0.05$) than those in MC across all landscapes types (Table 4). In contrast, there were no significant differences in MeHg stores between these watersheds among any of the landscape types, although MeHg stores in MC were marginally greater ($p = 0.058$) than those in FB. Stores of SOM were about eight-fold greater overall in FB than in MC, and these differences were highly significant ($p<0.001$) across all landscape types. These strong differences in SOM storage among the two watersheds suggest that THg and MeHg are more efficiently stored as a function of available SOM in MC than in FB. This suggestion is confirmed by noting that median values of THg/SOM were about three-fold greater in MC than in FB, and median values of MeHg/SOM were nearly four-fold greater in MC than in FB ($p<0.001$; Fig. 6).

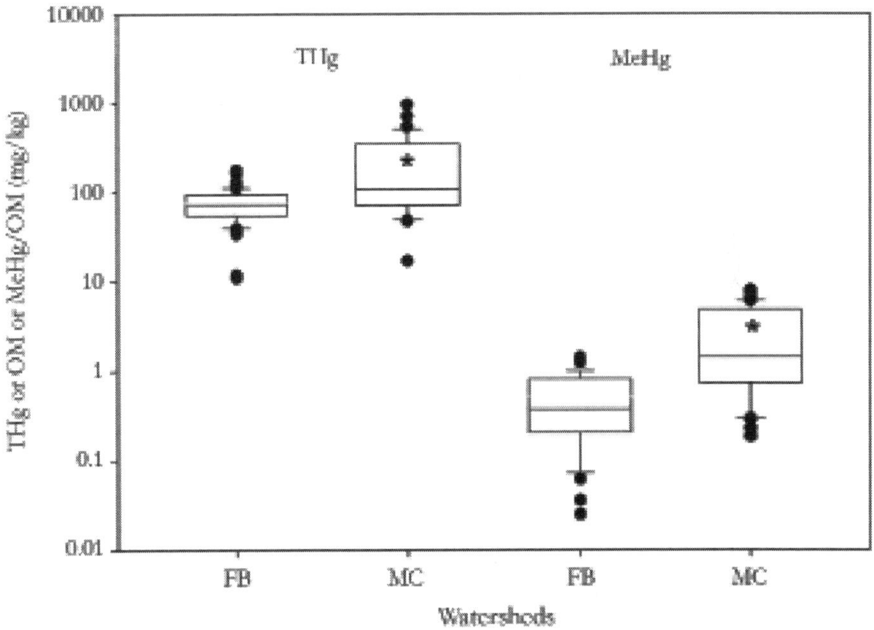

Figure 6. THg/OM and MeHg/OM based on estimates of storage in soil at each watershed. An asterisk in the box indicates the watershed with significantly greater values for each of the quantities shown.

Table 4. Median estimated areally-normalized storage of total Hg, methyl Hg, and organic matter to a depth of 40 cm in the Fishing Brook, NY and McTier Creek, SC watersheds

Site/Landscape Position	THg (pg/m²)	MeHg (mg/m²)	Organic Matter (kg/m²)
Fishing Brook			
All[1]	2570	8.4	38.8
Upland	2540	7.3	39.8
Riparian	3020	11.1	33.5
Wetland	2220	15.1	37.2
McTier Creek			
All[2]	690	10.3	4.7
Upland	650	11.0	2.6
Riparian	760	5.1	13.2
Wetland	980	11.1	12.7

1Represents 96.6% of watershed area excluding land cover not sampled in study (open water + developed land = 3.4%).

2Represents 79.3% of watershed area excluding land cover not sampled in study (open water +developed land + agricultural land = 20.7%)

The differences in soil stores of THg, MeHg, and SOM among landscape types within each watershed were less distinct than those identified between the watersheds (Table 4), but some differences were noted (values not shown). THg stores were statistically indistinguishable among landscape types in both watersheds. Similarly, there was no difference in MeHg soil stores among landscape types in MC. In contrast, a statistically significant difference in MeHg stores was identified in FB, with values decreasing in the order wetlands = riparian>uplands, according to pairwise comparisons. For SOM stores, no differences among landscape types were noted at FB, but these values differed across landscapes in MC, decreasing in the order uplands>wetlands = riparian.

DISCUSSION

Comparisons with Data from Other Studies

Hg concentrations and estimates of soil storage in these two watersheds are broadly consistent with values previously reported for the US and globally for rural sites where atmospheric deposition is the dominant input to the landscape [3], [6], [61], [62]. THg concentrations in soil from rural areas reported previously generally range from 50 to 300 ng/g, with greater values generally found in soil with high SOM levels, particularly in shallower soil horizons where SOM is concentrated [6], [7], [15], consistent with the results reported here. Fewer soil data have been reported for the southeastern US in locations similar to the MC watershed; however, two studies report THg concentrations in rural southeastern US sandy soil of about 10 ng/g [7] and 30 ng/g [63], broadly consistent with the generally lower median Hg values at MC (A = 60 ng/g, B = 30 ng/g) than those of FB (Oa = 200 ng/g, B = 85 ng/g).

Few studies have previously reported MeHg data in soil; however, some have reported MeHg/THg, with values as high as 5% [14], but most <1% [4], [6]. Here, we reported values of about 0.5% at FB and close to 1% at MC, consistent with most studies reviewed by Grigal [6] and Obrist [4]. Our data do not show a consistent pattern in MeHg/THg with soil depth across these two watersheds and the three landscape types sampled, but there is a pronounced tendency for the greatest values of MeHg and MeHg/THg to occur in riparian and wetland soil within each watershed. This pattern is consistent with those reported previously in these two watersheds and in other settings in which wetlands and riparian areas have the highest MeHg concentrations in soil, soil water, and shallow groundwater, and are the dominant source of MeHg to nearby surface waters [14], [25], [37].

The estimates of THg storage in FB soil of about 2500 μg/m^2 (to 40 cm depth) are within the range of most values reported previously in forested landscapes in the northeastern US [64], and lower than those reported in peat soil and peatlands [3], [6], [13], [65]. The FB THg soil stores are less than those in a nearby watershed in the western Adirondacks of New York [16], but these latter estimates included the

Oi and Oe horizons that were not sampled in FB, and also extended to a depth of 1 m, whereas THg stores were estimated only to a depth of 40 cm at FB. The estimated THg soil stores in MC of about 700 μg/m^2 are generally less than those reported in most previous studies, and at the low end of estimates of mean soil THg storage over large regions in the US and Canada [6], [66]. We are not aware of any previously published estimates of soil THg stores from uncontaminated sites in the US Coastal Plain other than that of Obrist [4] for a site near Gainesville, FL. Soil THg concentrations are about 200 ng/g in the Everglades [67], where the soil is rich in peat and not readily comparable to the sandy soil of the South Carolina Coastal Plain.

Comparisons between these Two Watersheds

The study watersheds have been previously identified as Hg "hot spots", and show little inter-watershed difference in THg and MeHg concentrations in surface waters, aquatic invertebrates, and fish [25], [26]. For example, median filtered THg and filtered MeHg concentrations were similarly about 2 ng/L and 0.1 ng/L, respectively at MC and FB during 2007–09 [68]. Although soil is the likely dominant immediate source of THg and MeHg to local streams and biota in these two watersheds (Hg in bottom sediment and direct atmospheric deposition are viewed as minor sources), there are large differences in THg concentrations and stores between these two locations, with about three-fold greater values in FB than in MC. As previously noted, however, there are large uncertainties in estimating chemical element storage in watersheds including accounting for rock content, which was not estimated in this study. Despite these uncertainties, we believe that the three-fold difference is greater than the uncertainties that arise in estimating Hg stores.

In contrast, differences in soil MeHg between these two watersheds show a more complex pattern. MeHg concentrations at FB were generally greater overall and in most landscape types than those at MC (though differences were not always statistically significant), but there was little difference in MeHg stores between the watersheds. Furthermore, MeHg/THg was generally greater in the MC watershed than in the FB watershed. These results are broadly consistent with previous studies that have shown increases in concentrations and stores of THg [4] and SOM [69] with increasing latitude in the US, but are inconsistent with previous observations that similarly showed soil MeHg concentrations and stores increasing with latitude [4].

The size of the soil THg pools relative to estimates of annual atmospheric deposition of Hg differs sharply at these two sites. The current soil THg pool is equivalent to about 400 years of stored Hg relative to the current annual atmospheric Hg deposition rate at Fishing Brook of 6.3 μg/m^2/yr, whereas the THg pool at McTier Creek is equivalent to about 70 years of stored Hg relative to current annual Hg deposition of 9.9 μg/m^2/yr [70], a more than 5-fold difference among these sites. In contrast, the same estimates for the soil MeHg pool suggests similar relative storage in these two watersheds (FB = 267 yrs, MC = 208 yrs, assuming MeHg deposition is 0.5% of THg deposition, [71]), though MeHg deposition rates are highly uncertain.

These findings suggest a conceptual model that includes the role of soil in the Hg cycle at these two locations. A much larger proportion of the atmospheric THg deposition that has historically fallen at the Adirondack site remains stored in watershed soil, consistent with the colder climate and larger SOM pool relative to the Coastal Plain. Thus, the soil in the FB watershed represents a large, persistent pool of THg in contrast to the soil of the MC watershed, which turns over more rapidly. Lower rates of annual evapotranspiration and higher annual streamflow in concert with the larger soil Hg pools in the Adirondack watershed imply the potential for greater annual stream THg yields at the more northern location, which is consistent with modeled estimates at these two sites during 2007–09 [70]. In the MC watershed, the rates of other loss processes such as gaseous emissions resulting from respiration of soil organic matter and volatilization of Hg^0 [2] are necessarily much greater than those in the Adirondack watershed to account for the much smaller THg pool in the Coastal Plain soil. Furthermore, the weak relation of THg to SOM in the Adirondack watershed soil with high organic matter content suggests that SOM composition is highly heterogeneous, consistent with the diverse specific ultra-violet absorbance values of DOC and relatively weak relation of DOC concentrations and filtered THg concentrations measured in Fishing Brook[72]. These results also suggest the likelihood of excess Hg binding capacity in this northern soil [8], whereas the retention of THg in the southern soil appears to be more strongly dependent on available SOM, which may be limiting retention. Greater net methylation efficiency is suggested at the southern Coastal Plain site because the pool sizes are similar among the two sites, the annual stream fluxes are greater at FB than at MC [70], and MeHg/THg is generally greater at MC. However, the role of demethylation rates, a potentially important factor not measured in this study, may also play an important role in mediating MeHg/THg ratios in soil [73]. Nonetheless, a greater proportion of the soil THg pool is likely converted to MeHg on an annual basis at the southern site, necessary to provide the similar MeHg concentrations in streams and aquatic biota observed among these two watersheds [25], [26]. We hypothesize that warmer soil temperatures in the southern Coastal Plain than in the Adirondacks is the primary driver of the apparent greater net methylation efficiency at MC. Other factors, however, such as differences in the vegetation and soil between these sites may also play a role.

A potentially important caveat to these results and the conclusions drawn is that soils were sampled only once in each watershed. Little is known from the literature about the extent to which soil concentrations and stores vary seasonally. Our observations of THg and MeHg in springs, streams, and piezometers indicate that aqueous concentrations can vary two- to three-fold throughout the year in these two watersheds [25], and a similar magnitude of variation is possible in soils, though the large size of the stores would be expected to dampen this variation relative to that in solution.

Differences across Landscape Types

Previously published findings from these two watersheds and nearby indicate that wetlands and riparian areas are major sources of THg and MeHg to local streams, and flow paths that transport Hg species from these landscape settings to streams are the dominant control on the spatial and temporal variations in concentrations and loads of these Hg species [25], [27], [37],[74]. When soil concentrations and stores of THg and MeHg are compared among landscape types, including low Hg-yielding uplands, the differences are not as striking as those observed in surface waters. There were no significant landscape differences in THg concentrations or stores in FB soil, but MeHg values were greater in wetland soil, particularly when compared with those of upland soil. MC had significantly greater THg and MeHg concentrations in wetlands than those in riparian areas and uplands, but the stores of these Hg species were not significantly different among the three landscape types. The data from the current study are consistent with a prominent role for factors other than soil THg and MeHg concentrations and stores as principal controls on landscape variation in Hg cycling rates to streams. These factors include soil drainage characteristics, proximity to the stream channel, and the high water table and fluctuating redox processes (anoxic to oxic) in the relatively flatter wetland and riparian terrain. In particular, differences in the hydrological and biogeochemical processes that affect the concentrations and relative fluxes of dissolved organic matter (DOM) as well as the relative aromatic vs. aliphatic character of this DOM appear to be the principal controls on landscape variation in THg and MeHg concentrations and loads in streams in these two geographic settings [25], [27], [72].

CONCLUSIONS

The results of this study show that two watersheds in different geomorphic and climatic regions with similar THg and MeHg concentrations in surface water and aquatic biota have greatly different concentrations and stores of THg in soil. THg concentrations and stores (to a depth of 40 cm) were about three-fold greater in the Adirondack watershed than those of the Coastal Plain watershed. Despite somewhat higher MeHg concentrations in the Adirondack soil, MeHg stores did not differ significantly among these two locations, although MeHg/THg was generally greater at the Coastal Plain site. These results are consistent with greater annual rates of net methylation and greater methylation efficiency in the soil of the warmer southern site. The greater soil stores of THg in the Adirondack watershed are consistent with an eight-fold greater store of SOM than that of the Coastal Plain watershed, though the relation of THg to SOM was weak in the Adirondacks, suggesting heterogeneous SOM and an excess of Hg-binding thiol groups relative to the net Hg supply. Furthermore, these results indicate that a greater proportion of historical atmospheric Hg deposition is stored in the Adirondack soil and that THg has about a five-fold greater soil turnover time than in the southern Coastal Plain soil. Comparison with known rates of atmospheric Hg deposition and stream export implies that gaseous Hg emissions from the Coastal Plain soil are necessarily much greater than those of Adirondack soil. These results raise questions about the ultimate fate of the Hg stored in the soil of these two sites. Past work has indicated residence times of several decades to several hundred years for atmospherically-deposited Hg, consistent with the results reported here. The overall greater store of THg in the Adirondacks suggests that declines in THg concentrations in stream water of this region will be more gradual and prolonged as future Hg deposition declines and comparatively sharper and less prolonged in Coastal Plain streams.

REFERENCES

1. Selin NE (2010) Global biogeochemical cycling of mercury: a review. Ann Rev Environ Resour 34: 43–63. doi: 10.1146/annurev.environ.051308.084314
2. Smith-Downey NV, Sunderland EM, Jacob DJ (2010) Anthropogenic impacts on global storage and emissions of mercury from terrestrial soils: insights from a new global model. J Geophys Res Biogeosci 115, G03008, doi:10.1029/2009JG001124.
3. Tipping E, Posakitt JM, Lawlor AJ, Wadsworth RA, Norris DA, et al. (2011) Mercury in United Kingdom topsoils; concentrations, pools, and Critical Limit exceedances. Environ Pollut 159: 3721–3729. doi: 10.1016/j.envpol.2011.07.011
4. Obrist D (2012) Mercury distribution across 14 U.S. forests. Part II: patterns of methyl mercury concentrations and areal mass of total and methyl mercury. Environ Sci Technol 46: 5921–5930. doi: 10.1021/es2045579
5. Alriksson A (2001) Regional variability of Cd, Hg, Pb and C concentrations in different horizons of Swedish forest soils. Water Air Soil Pollut Focus 1: 325–341.
6. Grigal DF (2003) Mercury sequestration in forest and soils: a review. J Environ Qual 32: 393–405.
7. Obrist D, Johnson DW, Lindberg SE, Luo Y, Hararuk O, et al. (2011) Mercury distribution across 14 U.S. forests. Part I: spatial patterns of concentrations in biomass, litter, and soils. Environ Sci Technol 45: 3974–3981. doi: 10.1021/es104384m
8. Skyllberg U, Qian J, Frech W, Xia K, Bleam WF (2003) Distribution of mercury, methyl mercury and organic sulphur species in

soil, soil solution and stream of a boreal forest catchment. Biogeochemistry 64: 53–76. doi: 10.1023/a:1024904502633

9. Ravichandran M (2004) Interactions between mercury and dissolved organic matter—a review. Chemosphere 55: 319–331. doi: 10.1016/j.chemosphere.2003.11.011
10. Wang W, Driscoll CT (1995) Patterns of total mercury concentrations in Onondaga Lake, New York. Environ Sci Technol 29: 2261–2266. doi: 10.1021/es00009a016
11. Gerbig CA, Ryan JN, Aiken GR (2012) The effects of dissolved organic matter on mercury biogeochemistry. In: Liu G, Cai Y, O'Driscoll NO editors. Environmental Chemistry and Toxicology of Mercury. New York: Wiley. 259–292.
12. Gabriel MC, Williamson DG (2004) Principal biogeochemical factors affecting the speciation and transport of mercury through the terrestrial environment. Environ Geochem Health 26: 421–434. doi: 10.1007/s10653-004-1308-0
13. Nater EA, Grigal DF (1992) Regional trends in mercury across the Great Lakes states, north central USA. Nature 358: 139–141. doi: 10.1038/358139a0
14. Schwesig D, Ilgen G, Matzner E (1999) Mercury and methylmercury in upland and wetland acid forest soils of a watershed in NE-Bavaria. Water Air Soil Pollut 113: 141–154.
15. Woodruff LG, Cannon WF, Eberl DD, Smith DB, Kilburn JE, et al. (2009) Continental-scale patterns in soil geochemistry and mineralogy: results from two transects across the United States and Canada. Appl Geochem 24: 1369–1381. doi: 10.1016/j.apgeochem.2009.04.009
16. Demers JD, Driscoll CT, Fahey TJ, Yavitt JB (2007) Mercury cycling in soil and litter in different forest types in the Adirondack region, New York, USA. Ecol Appl 17: 1341–1351. doi: 10.1890/06-1697.1
17. Chen X, Xia X, Wu S, Wang F, Guo X (2010) Mercury in urban soils with various types of land use in Beijing, China. Environ Pollut 158: 48–54. doi: 10.1016/j.envpol.2009.08.028
18. Woodruff LG, Cannon WF (2010) Immediate and long-term fire effects on total mercury in forests soils of northeastern Minnesota. Environ Sci Technol 44: 5371–5376. doi: 10.1021/es100544d

19. Fu X, Feng X, Zhu W, Rothenberg S, Yao H, et al. (2010) Elevated atmospheric deposition and dynamics of mercury in a remote upland forest of southwestern China. Environ Pollut 158: 2324–2333. doi: 10.1016/j.envpol.2010.01.032
20. Åkerblom S, Meili M, Bringmark L, Johansson K, Berggren D, et al. (2008) Partitioning of Hg between solid and dissolved organic matter in the humus layer of boreal forests. Water Air Soil Pollut 189: 239–252. doi: 10.1007/s11270-007-9571-1
21. Mitchell CPJ, Branfireun BA, Kolka RK (2008) Spatial characteristics of net methylmercury production hot spots in peatlands. Environ Sci Technol 42: 1010–1016. doi: 10.1021/es0704986
22. Wiener JG, Knights BC, Sandheinrich MB, Jeremiason JD, Brigham ME, et al. (2006) Mercury in soils, lakes, and fish in Voyageurs National Park (Minnesota): importance of atmospheric deposition and ecosystem factors. Environ Sci Technol 40: 6261–6268. doi: 10.1021/es060822h
23. Evers DC, Han Y-J, Driscoll CT, Kamman NC, Goodale MW, et al. (2007) Biological mercury hotspots in the northeastern United States and southeastern Canada. BioScience 57: 29–43. doi: 10.1641/b570107
24. Scudder BC, Chasar LC, Wentz DA, Bauch NJ, Brigham ME, et al.. (2009) Mercury in fish, bed sediment, and water from streams across the United States, 1998–2005. Reston, VA: US Geol Surv Sci-Invest Rpt 2009–5109. 86 p.
25. Bradley PM, Burns DA, Riva-Murray K, Brigham ME, Button DT, et al. (2011) Spatial and seasonal variability of dissolved methylmercury in two stream basins in the eastern United States. Environ Sci Technol 48: 2048–2055. doi: 10.1021/es103923j
26. Riva-Murray K, Chasar LC, Bradley PM, Burns DA, Brigham ME, et al. (2011) Spatial patterns of mercury in macroinvertebrates and fishes from streams of two contrasting forested landscapes in the eastern United States. Ecotoxicology 9: 1–13. doi: 10.1007/s10646-011-0719-9
27. Schelker J, Burns DA, Weiler M, Laudon H (2011) Hydrological mobilization of mercury and dissolved organic carbon in a snow-dominated forested watershed: conceptualization and modeling. J Geophys Res Biogeosci 116, G01002, doi:10.1029/2010JG001330.

28. Scudder Eikenberry BC, Riva-Murray K, Smith MJ, Bradley PM, Button DT, et al.. (2012) Environmental settings of streams sampled for mercury in New York and South Carolina, 2005–09. Reston, VA: US Geol Surv Open-File Rpt 2011–1318. 48 p.
29. McGee GG, Mitchell MJ, Leopold DJ, Raynal DJ, Mbila M (2007) Relationships among forest age, composition and elemental dynamics of Adirondack northern hardwood forests. J Torrey Botan Soc 134: 253–268. doi: 10.3159/1095-5674(2007)134[253:rafaca]2.0.co;2
30. LaPoint SD, Curren RP, Halasz SS, Barge JW, Spada DM, et al.. (2004) Wetlands effects database and GIS for the Adirondack Park. Ray Brook, NY: NY State Adirondack Park Agency. 26 p.
31. Kiernan BD, Hurd TD, Raynal DJ (2003) Abundance of Alnus incana ssp. rugosa in Adirondack Mountain shrub wetlands and its influence on inorganic nitrogen. Environ Pollut 123: 347–354. doi: 10.1016/s0269-7491(03)00023-x
32. Ashwal LD (1982) Mineralogy of mafic and Fe-Ti oxide-rich differentiates of the Marcy anorthosite massif, Adirondacks, New York. Am Mineral 67: 14–27.
33. Silverman MH, Krawiecki KJ (2006) Soil Survey of Hamilton County, New York. Washington DC: US Dept Agric, Nat Resourc Conserv Serv. 635 p.
34. Demers JD, Yavitt JB, Driscoll CT, Montesdeoca MR (2013) Legacy mercury and stoichiometry with C, N, and S in soil, pore water, and stream water across the upland-wetland interface: the influence of hydrogeologic setting, J Geophys Res Biogeosci. 118: 825–841. doi: 10.1002/jgrg.20066
35. Dahl TE, Dick J, Swords J, Wilen BO (2009) Data collection requirements and procedures for mapping wetland, deepwater and related habitats of the United States. Madison, WI: Div Habitat and Resource Conserv, National Stds and Support Team. 85 p.
36. Rogers VA (1985) Soil Survey of Aiken County area, South Carolina. Washington DC: US Dept Agric, Soil Conserv Serv. 135 p.
37. Burns DA, Riva-Murray K, Bradley PM, Aiken GR, Brigham ME (2012) Landscape controls on total and methyl Hg in the upper

Hudson River basin, New York, USA. J Geophys Res Biogeosci 117, G01034, doi:10.1029/2011JG001812.

38. Homer C, Hunag C, Yang L, Wylie B, Coan M (2004) Development of a 2001 National Land Cover Database for the United States. Photogram Engineer Remote Sensing 70: 829–840. doi: 10.14358/pers.70.7.829

39. NOWData – NOAA Online Weather Data. Available: http://nowdata.rcc-acis.org/BTV/pubACIS-results. Accessed 2014 Jan 20.

40. Park J-H, Mitchell MJ, Driscoll CT (2005) Winter-time climatic control on dissolved organic carbon export and surface water chemistry in an Adirondack forested watershed. Environ Sci Technol 39: 6993–6998. doi: 10.1021/es048301m

41. NOWData – NOAA Online Weather Data. Available: http://nowdata.rcc-acis.org/CAE/pubACIS-results. Accessed 2014 Jan 20.

42. USGS Water Data for the Nation. Available: http://waterdata.usgs.gov/nwis/sw. Accessed 2014 Jan 20.

43. Woodruff LG, Cannon WF, Knightes CD, Chapelle FH, Bradley PM, et al.. (2010) Total mercury, methylmercury, and selected elements in soils of the Fishing Brook watershed, Hamilton County, New York, and the McTier Creek Watershed, Aiken County, South Carolina, 2008. Reston, VA: US Geol Surv Data Ser Rpt 516. 18 p. Available: http://pubs.usgs.gov/ds/516. Accessed 2014 Jan 20.

44. Brown ZA, Curry KJ (2002) Total carbon by combustion. In: Surv Open-File Rpt Taggart JE Jr, editor. Analytical methods for chemical analysis of geologic and other materials. Denver, CO: US Geol. 02–223: 1–4 (Ch R)..

45. Brown ZA, Papp C, Brandt E, Aruscavage P (2002) Carbonate carbon by coulometric titration. In: Surv Open-File Rpt Taggart JE Jr, editor. Analytical methods for chemical analysis of geologic and other materials. Denver, CO: US Geol. 02–223: 1–6 (Ch S)..

46. Broadbent FE (1953) The soil organic fraction. Adv Agron 5: 153–183. doi: 10.1016/s0065-2113(08)60229-1

47. Smith DB, Woodruff LG, O'Leary RM, Cannon WF, Garrett RG, et al. (2009) Pilot studies for the North American Soil Geochemical

Landscapes Project – site selection, sampling protocols, analytical methods, and quality control protocols. Appl Geochem 24: 1357–1368. doi: 10.1016/j.apgeochem.2009.04.008

48. US EPA Method 7471B: Mercury in Solid or Semisolid Waste (Manual Cold Vapor Technique). Available: http://www.epa.gov/osw/hazard/testmethods/sw846/pdfs/7471b.pdf. Accessed 2104 Jan 20.
49. DeWild JF, Olund SD, Olson ML, Tate MT (2004) Methods for the preparation and analysis of solids and suspended solids for methylmercury, Reston, VA: U.S. Geol Surv Tech and Meth 5 A-7. 21 p.
50. DeSaules A (2012) Measurement instability and temporal bias in chemical soil monitoring: sources and control measures. Environ Monit Assess 184: 1487–1502. doi: 10.1007/s10661-011-1982-1
51. Harrison RB, Adams AB, Licata C, Flaming B, Wagoner GL, et al. (2003) Quantifying deep-soil and coarse-soil fractions: avoiding sampling bias. Soil Sci Soc Am J 67: 1602–1606. doi: 10.2136/sssaj2003.1602
52. Adams WA (1973) The effect of organic matter on the bulk and true densities of some uncultivated podzolic soils. J Soil Sci 24: 10–17. doi: 10.1111/j.1365-2389.1973.tb00737.x
53. Post WM, Kwan KC (2000) Soil carbon sequestration and land-use change: processes and potential. Global Change Biol 6: 317–28. doi: 10.1046/j.1365-2486.2000.00308.x
54. De Vos B, Van Meirvenne M, Quataert P, Deckers J, Muys B (2005) Predictive quality of pedotransfer functions for estimating bulk density of forest soils. Soil Sci Soc Am J 69: 500–510. doi: 10.2136/sssaj2005.0500
55. Markewich HW, Pavich MJ, Buell GR (1990) Contrasting soils and landscapes of the Piedmont and Coastal Plain, eastern Unites States. Geomorphology 3: 417–447. doi: 10.1016/0169-555x(90)90015-i
56. Duffera M, White JG, Weisz R (2007) Spatial variability of southeastern U.S. Coastal Plain soil physical properties: implications for site-specific management. Geoderma 137: 327–339. doi: 10.1016/j.geoderma.2006.08.018

57. Federer CA, Trucotte DE, Smith CT (1993) The organic fraction – bulk density relationship and the expression of nutrient content in forest soils. Can J Forest Res 23: 1026–1032. doi: 10.1139/x93-131
58. Mann HB, Whitney DR (1947) On a test of whether one of two random variables is stochastically larger than the other. Ann Math Stat 18: 50–60. doi: 10.1214/aoms/1177730491
59. Kruskal WH, Wallis WA (1952) Use of ranks in one-criterion variance analysis. J Am Stat Assoc 47: 583–621. doi: 10.1080/01621459.1952.10483441 View ArticlePubMed/NCBI
60. Helsel DR (2005) Nondetects and data analysis. Hoboken, NJ: Wiley-Interscience. 250 p.
61. Rundgren S, Ruhling Å, Schlütter K, Tyler G (1992) Mercury in soil–distribution, speciation, and biological effects. Copenhagen: Nordic Council of Ministers. 89 p.
62. Burt R, Wilson MA, Mays MD, Lee CW (2003) Major and trace elements of selected pedons in the USA. J Environ Qual 32: 2109–2121. doi: 10.2134/jeq2003.2109
63. Aelion CM, Davis HT, McDermott S, Lawson AB (2009) Soil metal concentrations and toxicity: Associations with distances to industrial facilities and implications for human health. Sci Total Environ 407: 2216–2223. doi: 10.1016/j.scitotenv.2008.11.033
64. Richardson JB, Friedland AJ, Engerbretson TR, Kaste JM, Jackson BP (2013) Spatial and vertical distribution of mercury in upland forest soils across the northeastern United States. Environ Pollut 182: 127–134. doi: 10.1016/j.envpol.2013.07.011
65. Lindqvist O, Johansson K, Aastrup M, Andersson A, Bringmark L, et al. (1994) Mercury in the Swedish environment–recent research on causes, consequences and corrective measures. Water Air Soil Pollut 55: 1–253. doi: 10.1007/bf00542429
66. Richardson GM, Mitchell IA, Mah-Apulson M, Hackbarth T, Garrett RG (2003) Natural emissions of mercury to the atmosphere in Canada. Environ Rev 11: 17–36. doi: 10.1139/a03-004
67. Arfstrom C, MacFarlane AW, Jones RD (2000) Distributions of mercury and phosphorus in Everglades soils from Water Conservation Area 3A, Florida, U.S.A. Water Air Soil Pollut. 121: 133–159. doi: 10.1023/a:1005214908516

68. Bradley PM, Journey CA, Brigham ME, Burns DA, Button DT, et al. (2013) Intra- and inter-basin mercury comparisons: Importance of basin scale and time-weighted methylmercury estimates. Environ Pollut 172: 42–52. doi: 10.1016/j.envpol.2012.08.008
69. Guo Y, Gong P, Amundson R, Yu Q (2006) Analysis of factors controlling soil carbon in the conterminous United States. Soil Sci Soc Am J 70: 601–612. doi: 10.2136/sssaj2005.0163
70. Journey CA, Burns DA, Riva-Murray K, Brigham ME, Button DT, et al.. (2012) Fluvial transport of mercury, organic carbon, suspended sediment, and selected major ions in contrasting stream basins in South Carolina and New York: October 2004–September 2009. Columbia, SC: U.S. Geol Surv Sci-Invest Rpt 2012–5173. 128 p.
71. Bushey JT, Nallana AG, Montesdeoca MR Driscoll CT (2008) Mercury dynamics of a northern hardwood canopy. Atmos Environ 42: 6905–6914. doi: 10.1016/j.atmosenv.2009.02.023
72. Burns DA, Aiken GR, Bradley PM, Journey CA, Schelker J (2013) Specific ultra-violet absorbance as an indicator of mercury sources in an Adirondack river basin. Biogeochemistry 113: 451–466. doi: 10.1007/s10533-012-9773-5
73. Tjerngren I, Karlsson T, Björn E, Skyllberg U (2012) Potential Hg methylation and MeHg demethylation rates related to the nutrient status of different boreal wetlands. Biogeochemistry 108: 335–350. doi: 10.1007/s10533-011-9603-1
74. Shanley JB, Schuster PF, Reddy MM, Roth DE, Taylor HE, et al.. (2002) Mercury on the move during snowmelt in Vermont, Eos Trans Am Geophys Union 83: 45, 47–48.

Chapter 7

Do Estimates of Water Productivity Enhance Understanding of Farm-Level Water Management?

Dennis Wichelns

P.O. Box 2629, Bloomington, IN 47402, USA;

ABSTRACT

Estimates of water productivity are appearing with increasing frequency in the literature pertaining to agronomy, water management, and water policy. Some authors report such estimates as one of the outcome variables of experiment station studies, while others calculate water productivities when comparing regional crop production information. Many authors suggest or imply that higher values of water productivity are needed to ensure that future food production goals are achieved.

Yet maximizing water productivity might not be consistent with farm-level goals or with societal objectives regarding water allocation and management. Farmers in both rainfed and irrigated settings must address a complex set of issues pertaining to risk, uncertainty, prices, and opportunity costs, when selecting activities and determining optimal strategies. It is not clear that farmers in either setting will or should choose to maximize water productivity. Upon examining water productivity, both conceptually and empirically, using published versions of crop production functions, I conclude that estimates of water productivity contain too little information to enhance understanding of farm-level water management.

MOTIVATION

Many authors have called for improvements in water use efficiency or increases in water productivity in rainfed and irrigated settings, with the goal of increasing agricultural output to meet future food demands [1,2,3,4,5]. Some of the phrases that appear often in the agricultural and water management literature include the need to produce "more food with less water" and to generate "more crop per drop" [6,7,8,9]. The motivating rationale for such phrases is clear and legitimate. We must produce more food in future to support a larger global population. The competition for water is increasing in many sectors and water is a critical input in agriculture. Thus we must use the resource wisely to ensure that future generations will have access to reliable and affordable food supplies [10,11,12,13,14,15].

The notion of increasing crop water productivity is compelling at first glance, precisely because it suggests we need to generate more agricultural output per unit of the water input. Yet, focusing on water alone might not be sufficient to fully understand current crop production activities or to gain insight regarding the best measures for increasing crop yields or agricultural incomes in future.

Water is just one of many inputs in agriculture. Plant and crop responses to water generally depend on the availability of other inputs, such as nutrients, sunshine, and management effort [2,16,17,18]. Farmers seeking to increase crop yields or to achieve some other objective must optimize over an array of agricultural inputs, while considering also their market opportunities. In addition, farmers in both rainfed and irrigated settings must manage the risk and uncertainty inherent in their production activities, to determine optimal choices.

My goal in this paper is to consider whether estimates of crop water productivity are helpful in gaining a better understanding of water management in agriculture, and whether they are useful in evaluating crop production opportunities at the farm level. My primary concern is that estimates of crop water productivity lack the depth and breadth of information needed to understand a crop's response to water and a farmer's response to the natural, policy, and market environments in which he or she operates. By considering only one of many inputs, measures of crop water productivity omit much of the information that farmers consider and many of the issues that influence crop response to rainfall, irrigation, and cultural practices.

I hope also to motivate greater consideration of the risk and uncertainty many smallholder farmers face when making decisions regarding water and other inputs. In many developing countries, small-scale farmers struggle to gain access to irrigation water, to increase crop and livestock output, and enhance their livelihoods [19,20,21,22,23]. They operate in an environment of considerable risk and uncertainty. It is not clear that they could or should endeavor to maximize water productivity. Both farmers and policy makers must incorporate risk and uncertainty in their analyses of production opportunities and in decisions regarding natural resources.

I begin by describing measures of water productivity in the context of crop water production functions, with the goal of distinguishing between average and marginal measures of crop responses to farm inputs. Using published crop water production functions, I show that farmers seeking to maximize water productivity would generate less yield and smaller net income than if they sought to maximize net revenue. Thus, it is not clear that farmers would pursue the goal of maximizing water productivity, as proposed or implied by many authors. In cases in which water is limiting, relative to land, the profit-maximizing solution will be the same as that which maximizes the average productivity of water, but that is a special case result of the more general objective of maximizing net revenue.

I discuss also the random nature of measures of water productivity. Both the numerator and denominator of the estimated ratios contain deterministic and stochastic components. Crop yields and evapotranspiration are influenced by farmer decisions and input use, and also by random effects due to weather, pests, and disease. Thus, estimates of crop water productivity are random variables that represent point estimates from underlying probability distributions.

I also depict the challenge of comparing any two estimates of water productivity, whether they pertain to different settings or different times. The information contained in point estimates of water productivity is not sufficient for declaring that observations reflecting higher measures of water productivity should be preferred over those with lower measures. It is not clear that we gain helpful information or insight by comparing water productivities across locations or over time.

AVERAGE *VS.* MARGINAL PRODUCTIVITY

Crop water productivity, as defined by most authors, is the ratio of crop yield or crop value, to a selected measure of water consumed, applied, or evaporated in the process of growing a crop [2,24]. As such, the ratio represents the average productivity of the input, rather than the incremental productivity. For example, when the ratio of interest is the water productivity of applied water, the ratio often is defined as:

$$WP_{AW} = \text{Crop yield (tons/ha)/Applied water } (m^3/\text{ha}) \qquad (1)$$

If the crop yield is 4 tons per ha and the applied water is 8000 m^3 per ha, then the crop water productivity is 0.0005 tons per m^3. In addition to neglecting consideration of other inputs, this measure of average productivity is not sufficient for determining whether the application rate of 8000 m^3 per ha is optimal from the farm-level or societal perspective. The question of optimality can be addressed only by considering the marginal (or incremental) productivity of water, in comparison with its marginal cost. If the marginal productivity of water is 0.004 tons per m^3, and the price of the crop is $1,000 per ton, then the decision to apply the last m^3 of water is sensible, provided the marginal cost of water is less than or equal to $4.00 per m^3.

The average and marginal productivities of water can be depicted readily in the context of a crop water production function. This context is helpful also for describing why average productivities are not sufficient for determining optimal levels of water use. A typical crop water production function resembles the following diagram, which describes crop yield as a function of applied water.

The slope of a ray from the origin, through any point on the production function, depicts the average productivity of that amount of applied water. For example, as depicted in Figure 1, the average productivity of AW_1 (m^3 per ha) is calculated by dividing Y_1 by AW_1. That calculation is precisely the slope of the ray from the origin through the production function. Thus, the slope of the ray also represents the value of WP_{AW} in this example, or the crop water productivity when applying AW_1 units of water.

Beginning from AW_0, the value of WP_{AW} will increase with larger volumes of applied water, to a point, before beginning to decline. The point at which the value of WP_{AW} reaches its maximum is precisely the point at which a ray from the origin is just tangent to the production function. This occurs where the amount of applied water is AW_{MAX} (Figure 2). Beyond this level of applied water, the estimated water productivity will decline throughout the remaining range of applied water values. One can infer from the diagram that the point of maximum water productivity might be very different from the point of maximum crop yield. It might also be quite different from the point of maximum net revenue.

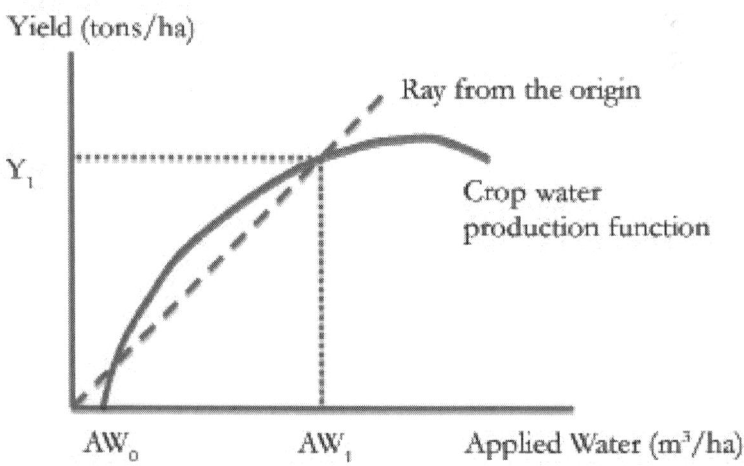

Figure 1. Depicting the water productivity of applied water in a typical crop water production function.

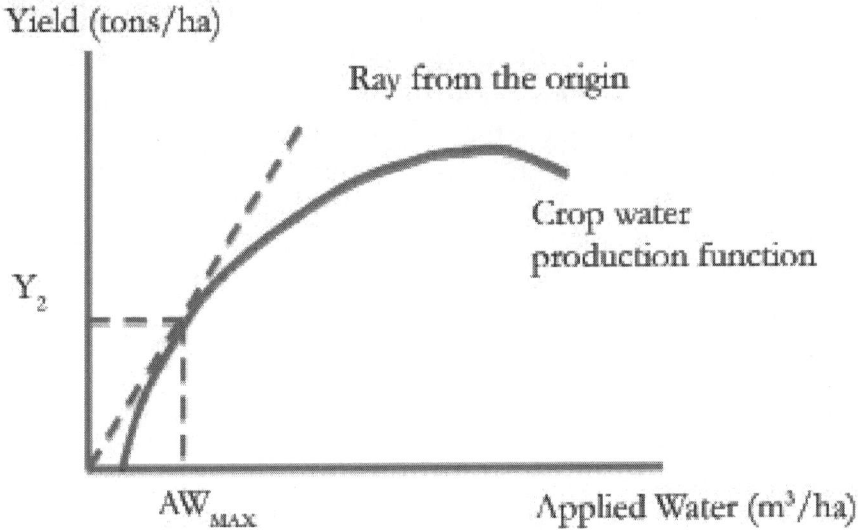

Figure 2. Depicting the maximum value of the water productivity of applied water.

The profit-maximizing amount of applied water, AW_{NR}, is found where marginal revenue is equal to marginal cost. To determine this amount, we first transform the crop water production function into a crop revenue function by multiplying by crop price. The shape of the crop revenue function will be the same as the crop water production function. We can depict the total cost of water as a straight line from the origin, provided that the unit price of water is constant. As shown in Figure 3, the profit-maximizing amount of applied water is larger than the volume that would maximize water productivity, and yet smaller than the volume that would maximize crop yield. This will be the case for many crops in a range of settings. Indeed, there is no particular importance attached to the water volume that maximizes water productivity, or equivalently, the average productivity of water.

It is also worth noting that for a crop water production function that arises from the origin or has a positive intercept on the vertical axis, water productivity will decline throughout the entire range of volumes of applied water. In that case, it is easy to determine that maximizing water productivity would not be the appropriate optimizing criterion, as it would generate the recommendation of using zero applied water. Such a production function would pertain in cases in which rainfall is sufficient to generate some yield, but farmers apply supplemental irrigation water to enhance crop production. The water productivity of the applied irrigation water would decline throughout the range of applied water volumes.

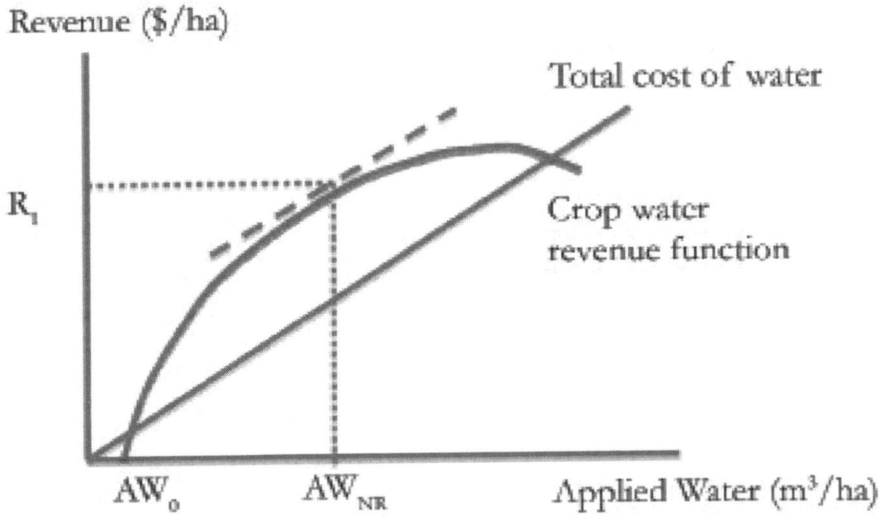

Figure 3. Depicting the profit-maximizing amount of applied water.

EMPIRICAL EXAMPLES

Several authors provide empirical examples of production functions that are useful in demonstrating the difference in outcomes when maximizing water productivity or maximizing crop yield or net revenue. Igbadun *et al.* [25] examined onion production in Nigeria. They presented the following estimated crop water production function in which the yield of onion bulbs is expressed as a function of seasonal applied water:

$$\text{Yield (tons/ha)} = -14.88 + 0.131\ AW - 0.0001\ AW^2 \qquad (2)$$

The applied water variable, AW, is expressed in mm.

Given this empirical production function, farmers could maximize water productivity by applying 390 mm of irrigation water. The yield of onion bulbs would be 21 tons per ha and the water productivity would be 54 kg of onion bulbs per mm of applied water (Table 1). As noted above, this is equivalent to the average productivity of applied water.

Farmers could achieve the maximum yield of 28 tons per ha by applying 660 mm of irrigation water. The crop water productivity would decline to 42 kg per mm at maximum yield, but the farmers would be producing an additional 7 tons of onion bulbs per ha. Lacking a cost function, it is not possible to determine the profit-maximizing volume of applied water. However, we can expect that if onion production is profitable, the optimal volume of applied water will be between 390 mm, where crop water productivity is maximum, and 660 mm, where yield is maximum and the marginal productivity of water is zero.

Zhang and Oweis 26 examine durum wheat production in Syria, where farmers rely partly on rainfall and partly on supplemental irrigation. Their estimated production is the following:

$$\text{Yield (tons/ha)} = -5.8556 + 0.0329\,(AW + P) - 0.00002164\,(AW + P)^2 \quad (3)$$

Both the supplemental irrigation (AW) and precipitation (P) are expressed in mm.

We determine the volumes of water that will maximize crop water productivity, yield, and net revenue, after choosing a seasonal precipitation level of 250 mm. In such a season, farmers can maximize crop water productivity by applying an additional 270 mm of irrigation water (Table 1). The yield would be 5.4 tons of wheat per ha, and the crop water productivity would be 2.00 kg of wheat per m^3 of applied water. The maximum wheat yield of 6.65 tons per ha can be achieved with supplemental irrigation of 510 mm. Crop water productivity would decline to 1.30 kg per m^3, but farmers would be producing an additional 1.25 tons of wheat per ha.

Table 1. Examples of water levels that maximize water productivity, net revenue, and yields, from empirical crop water production functions

Citation	**Crop and Location**	**Applied Water (mm)**	**Yield (t/ha)**	**CWP (kg/m^3)**	**Net Rev. ($/ha)**
Igbadun *et al.*, 2012 25	Onions, Nigeria	$Y = -14.88 + 0.131\ AW - 0.0001\ AW^2$			
	Max. CWP	390	21.0	5.38	1709
	Max. Net Rev	n/a	n/a	n/a	n/a
	Max. Yield	660	28.0	4.24	2141
Zhang and Oweis, 1999 26	Wheat, Syria	$Y = -5.8556 + 0.0329\ (AW + P) - 0.00002164\ (AW + P)^2$			
	Max. CWP	270	5.40	2.00	866
	Max. Net Rev	454	6.58	1.45	1048
	Max. Yield	510	6.65	1.30	1032
Zhang *et al.*, 1999 27	Wheat, Syria	$Y = 307 + 1.839\ AW - 0.00402\ AW^2$			
	Max. CWP	0	3.07	n/a	491
	Max. Net Rev	195	5.13	2.63	737
	Max. Yield	230	5.17	2.25	729

Notes: CWP refers to crop water productivity, which is defined as yield divided by applied water, and expressed in kg per m3. In the example of Zhang and Oweis (1999) 27, we assume seasonal rainfall of 250 mm

Zhang and Oweis 26 provide cost information that enables us to determine the profit-maximizing amount of supplemental irrigation. In a season with 250 mm of rainfall and a wheat price of $0.25 per kg, farmers would maximize profit by applying 454 mm or supplemental irrigation, generating a wheat yield of 6.58 tons per ha. Crop water productivity would be 1.45 kg per m^3. As expected, this profit-maximizing result lies between the volumes of water that maximize water productivity and crop yield. The difference in profit between the water applications that maximize crop water productivity and net revenue is $182 per ha.

Zhang *et al.* [27] provide an empirical production function for winter wheat in China, which has a positive intercept on the yield axis. Given this characteristic, as noted above, crop water productivity will decline throughout all values of applied water. The estimated production is the following:

$$\text{Yield (g/m}^2) = 307 + 1.839\ AW - 0.00402\ AW^2 \quad (4)$$

Farmers wishing to maximize crop water productivity would abstain from irrigating, thus obtaining a yield of 307 g per m², or 3.07 tons per ha. Crop water productivity is not defined at the value of zero mm of irrigation water. Yet the pertinent point is that water productivity declines for any positive value of applied water. Farmers could achieve the maximum yield of 5.2 tons per ha by applying 230 mm of irrigation water (Table 1). Farmers wishing to maximize net revenue would apply 195 mm of irrigation water, thus generating a yield of 5.1 tons per ha and obtaining a profit of $737 per ha. Here again, the profit-maximizing volume of applied water lies between the volumes that maximize water productivity and crop yield.

A Special Case

The above examples demonstrate empirically that farmers applying irrigation water would not have an interest in maximizing water productivity, as that strategy would not maximize yields or net revenue. Yet there is a special case in which the goal of maximizing net revenue generates a strategy that involves maximizing the average productivity of water. In particular, when water is limiting, relative to land, the profit maximizing solution will be the same as that which maximizes water productivity.

We can demonstrate this result using the empirical production function for maize, presented by Zand-Parsa and Sepaskhah [28], in which maize yield is a function of nitrogen and irrigation water:

$$\text{Yield (tons/ha)} = a_0 + a_1 \times W + a_2 \times W^2 + a_3 \times (N + N_r) + a_4 \times (N + N_r)^2 + a_5 \times (W)(N + N_r) + a_6 \times (W)(N + N_r)^2 + a_7 \times (W^2)(N + N_r) + a_8 \times (W^2)(N + N_r)^2 \quad (5)$$

In this equation, W represents the depth of irrigation water applied (m), N is nitrogen applied (kg per ha), and N_r is residual nitrogen in the soil before planting (kg per ha). The empirical values of the coefficients a_0 through a_8 are −1.862, 6.0386, −3.580, −0.178, 0.0005, 0.5733, −0.00148, −0.3021, and 0.000806. The assumed value of N_r is 54.6 kg per ha. The prices in the original analysis of Zand-Parsa and Sepaskhah 28 for output, water, and nitrogen are $200 per ton, $31 per m ($0.003

per m³), and \$0.13 per kg, respectively. In the standard case, in which water is plentiful, but the farmer's land is limited, the profit maximizing levels of water and nitrogen are 990 mm and 212 kg per ha (Table 2). As expected, the net revenue of \$2,556 per ha is larger than the net revenue that would be achieved if the farmer chose to maximize water productivity. Given that goal, the farmer would apply 723 mm of irrigation water and 207 kg per ha of nitrogen, while earning net revenue of \$2,152 per ha.

Table 2. Optimizing water and nitrogen applications for maize, when either land or water is the limiting input

Citation	Crop and Location	Applied Water (mm)	Applied N (kg/ha)	Yield (t/ha)	CWP (tons/m)	Net Rev. ($/ha)
28	Maize, Iran	Yield (t/ha) = $a_0 + a_1 \times W + a_2 \times W^2 + a_3 \times (N + N_r) + a_4 \times (N + N_r)^2 + a_5 \times (W)(N + N_r) + a_6 \times (W)(N + N_r)^2 + a_7 \times (W^2)(N + N_r) + a_8 \times (W^2)(N + N_r)^2$				
Land is limiting	Max. CWP:	723	207	11.0	15.2	2,152
	Max. Net Rev:	990	212	13.1	13.2	2,556
	Max. Yield:	1,000	212	13.1	13.1	2,555
Water is limiting	Max. CWP:	723	207	11.0	15.2	2,152
	Max. Net Rev:	723	207	11.0	15.2	2,152
	Uniform water:	567	254	8.2	14.5	1,588

Notes: CWP refers to crop water productivity, which is defined as yield divided by applied water, and expressed in tons per m3. CWP refers to crop water productivity, which is defined as yield divided by applied water, and expressed in tons per m3. In the uniform water example, we assume the famer has the equivalent of 1.7 m of irrigation water available and 3.0 ha of land that are available for cultivation. The optimal amount of land to plant and irrigate is 2.35 ha, while the optimal amounts of applied water and nitrogen are those shown in the table.

When water is limiting, relative to land, the amount of irrigation water that maximizes profit is the same as that which maximizes water productivity. Thus, the farmer would choose to apply 723 mm of irrigation water, while leaving some land idle. For example, suppose a farmer has 3.0 ha of land available, but only the equivalent of 1.7 m of irrigation water available. The farmer could plant all 3 ha in maize and apply 567 mm of irrigation water on those planted hectares. If the farmer also applies 254 kg/ha of nitrogen, he or she will achieve a yield of 8.2 tons/ha, while generating a net revenue of \$1,588 per ha, or $4,764 per season for the farm (Table 2). By contrast, if the farmer plants only 2.35 ha in maize and applies 723 mm of water and 207 kg/ha of nitrogen on those hectares, he or she will generate a net revenue of $2,152 per cultivated ha, or $5,057 for the farm.

This empirical example applies to situations in which water is limiting, relative to land. In such cases, the profit maximizing amount of land in cultivation can be determined by dividing the amount of water available by the amount of water per hectare that maximizes average productivity. In this example, the area of 2.35 ha can be determined by dividing the 1.7 m available by the 0.723 m at which water productivity is maximized. Thus, the strategy of maximizing water productivity in this instance is a special case result of the broader economic objective of maximizing profit.

THE PRODUCTIVITY OF WATER EVAPOTRANSPIRED

The production functions examined above depict the empirical relationship between crop yield and applied water. The curvature of each function reflects the diminishing and eventually negative incremental returns to applied water, as some of the water is not used productively by the plants. Functions describing crop yield or biomass as an outcome of evapotranspiration (ET) are essentially linear, for a given production setting [24,29,30]. Thus, additional evapotranspiration generates additional yield at a rate that is essentially constant. Despite this constant rate of transformation, estimated values of water productivity can increase, decline, or remain the same, as the volume of water evapotranspired increases.

Linear segments depicting crop yield as a function of evapotranspiration cannot intercept the vertical axis, as it is not possible to achieve a positive yield with no ET. Yet one can project where the vertical axis intercept would occur, if the linear segment were extended to the axis. Such an exercise is helpful in determining whether water productivity will increase or decrease at higher levels of ET for a given crop-ET relationship.

For example, if the linear segment depicting crop yield as a function of ET projects a positive vertical axis intercept, the empirical value of crop water productivity will decline as ET increases. This can be verified by noting that the slope of a ray from the origin through a point on the production function diminishes at higher levels of ET (Figure 4). By analogy, if the linear segment depicting crop yield as a function of ET projects a negative vertical axis intercept, the empirical value of crop water productivity will increase as ET increases (Figure 5). If the linear segment happens to be coincident with a 45-degree ray from the origin, the value of crop water productivity will be the same for all amounts of evapotranspiration.

Given that water productivity can increase or decrease monotonically along the linear segment depicting crop yield as a function of ET, it is not clear that the objective of increasing or maximizing water productivity, with respect to ET, is pertinent. If the empirical relationship between yield and ET resembles the one in Figure 4, water productivity will be maximized at the smallest amount of ET. Yet that level would also generate the smallest yield. By contrast, if the empirical relationship resembles Figure 5, there is no apparent upper bound to the value of water productivity. Higher measures of water productivity can be achieved at ever higher levels of ET. Such a relationship is not helpful in determining the optimal amount of evapotranspiration at the farm level.

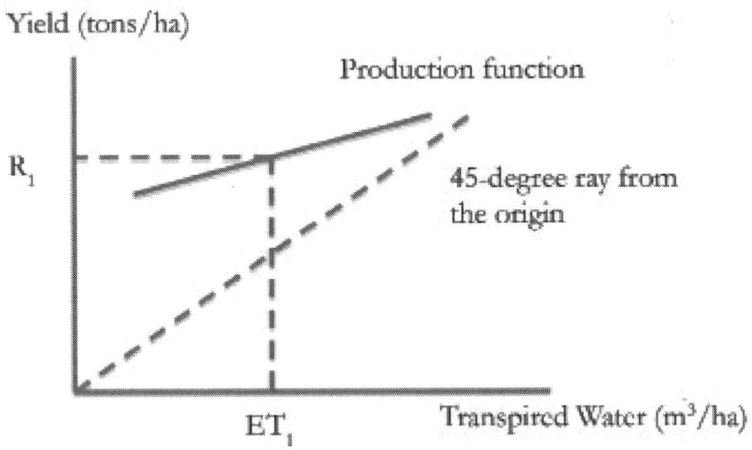

Figure 4. Depicting crop water productivity of evapotranspired (ET) water, when the linear segment projects a positive vertical axis intercept.

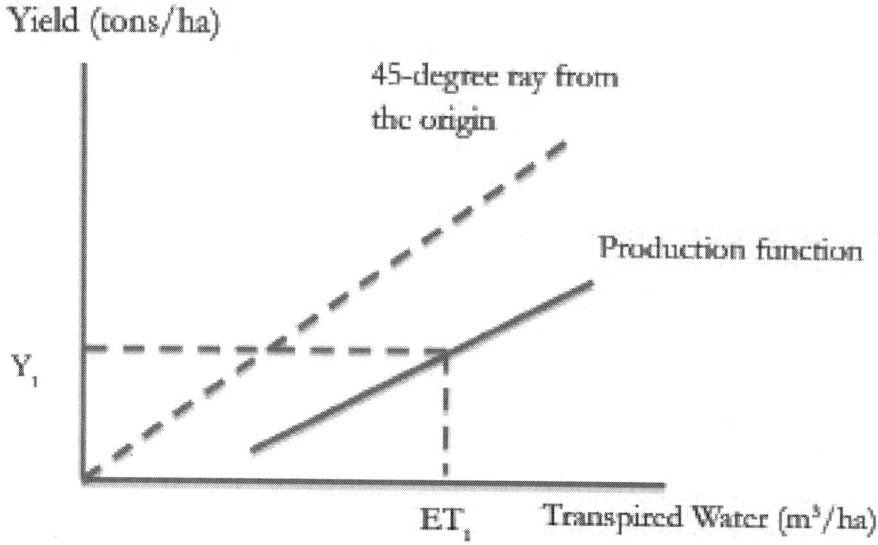

Figure 5. Depicting crop water productivity of evapotranspired water, when the linear segment projects a negative vertical axis intercept.

Empirical examples of linear relationships involving crop yield and ET are presented in Table 3. Three of these examples project a negative vertical intercept, while one projects a positive intercept. The R^2 values for these estimated relationships are higher than some others that appear in the literature, in part because the equations are estimated using relatively small numbers of observations. In larger scale studies, the scatter of observations is substantial, such that the measures of R^2 are much smaller. Depictions of linear yield-ET relationships for wheat, rice, cotton, and maize in Zwart and Bastiaanssen [24] carry R^2 values ranging from 0.09 to 0.39. Those authors acknowledge that the empirical description of crop yield and ET in any setting is determined by "many factors that influence the soil-plant-water relationship."

Table 3. Examples of linear crop production functions expressing yield as a function of evapotranspiration (ET)

Citation	Crop and Location	Estimated Production Function and R^2
Zhang and Oweis, 1999 26	Bread wheat, Syria	$Y = -2.4969 + 0.016$ ET $R^2 = 0.68$
		Yield is in tons per ha, ET is in mm.
Zhang and Oweis, 1999 26	Durum wheat, Syria	$Y = -0.1093 + 0.116$ ET $R^2 = 0.68$
		Yield is in tons per ha, ET is in mm.
Karam *et al.*, 2009 31	Wheat, Lebanon	$Y = 1364.6 + 5.0601$ ET $R^2 = 0.60$
		Yield is in kg per ha, ET is in mm.
Igbadun *et al.*, 2012 25	Onions, Nigeria	$Y = -0.4956 + 0.052$ ET $R^2 = 0.87$
		Yield is in tons per ha, ET is in mm.

Notes: Estimates of crop water productivity will diminish with increasing values of ET for the estimated functions that project a positive intercept. Measures of crop water productivity will increase with increasing values of ET for the estimated functions that project a negative intercept.

ESTIMATES OF WATER PRODUCTIVITY ARE RANDOM VARIABLES

Crop yield is a random variable in both rainfed and irrigated settings. So too is soil moisture, at any point in time, or as measured cumulatively throughout a season. Thus, the ratio representing water productivity, whether defined as biomass, crop yield, or revenue per unit of water applied or consumed, is essentially a ratio involving two random variables. Farmers certainly can influence the yields they achieve each season through their choices and timing of crop inputs, such as water, nutrients, pesticides, and labor, but the yields they obtain are random draws from a probability distribution. The same is true for soil moisture. Farmers with access to irrigation can modify soil moisture to some degree, but they cannot determine with certainty the amount of moisture available to a plant or the amount of moisture the plant takes up. Nonetheless, farmers with irrigation can notably influence plant growth and crop yield. Farmers lacking irrigation can manage soil moisture through their choices of crops and varieties, soil preparation, planting dates, plant densities, and cultural practices [32,33,34,35,] but they generally cannot supplement soil moisture autonomously. For both groups of farmers, the amount of soil moisture available at any point in time, and the uptake of soil moisture by plants, are random draws from probability distributions.

Given that measures of water productivity are ratios of two random variables, it is not clear how much influence farmers can exert on the empirical values observed on farms or computed across agricultural areas. Thus, the seemingly compelling goal of increasing water productivity might have limited relevance in most agricultural settings. Rather than attempting to increase water productivity, farmers might endeavor instead to reduce the variance in crop yields and soil moisture, while also increasing the mean values of crop yields, over time, in part by gaining greater influence over the seasonal pattern of soil moisture. Many farmers, particularly those in rainfed settings, likely already adopt such a management objective, either explicitly or implicitly. It is not possible to farm successfully in rainfed settings without paying close attention to soil moisture conditions, and managing accordingly.

Farmers in rainfed and partially irrigated settings can reduce the variance and increase the mean of their crop yields by investing in measures that modify soil moisture availability [36,37]. Farmers who capture rainwater in a small pond for irrigation, or pump groundwater from a shallow aquifer, can manage the amount of soil moisture available during crop growth and reproductive stages. They have the options of stimulating rapid growth of the plant canopy early in the season to reduce evaporation, and applying supplemental water during particularly dry stretches of a crop production cycle [38].

The actions taken by farmers to modify soil moisture might cause the observed value of water productivity to increase or decrease, depending on the final measured values of crop yields and water applied or consumed. Given the random nature of both outcomes, and interactions with other inputs, it is not possible to know in advance the eventual measure of water productivity. Yet the farmers will have made better use of the water within their control, by modifying the pattern of soil moisture to match crop water demands, and by striving to maximize transpiration while minimizing non-beneficial evaporation.

COMPARING ESTIMATES OF WATER PRODUCTIVITY

Many authors compare estimates of water productivity, over time or in selected locations. Examples include the comprehensive review by Zwart and Bastiaanssen [24] and smaller, regional reviews by Hussain *et al.* [39] and Yan and Wu [40]. One goal of such studies is to highlight differences in crop water productivities that might provide insight regarding policy measures or technological interventions that might improve water management and thus lead to greater output per unit of water used in agriculture. This goal is compelling, yet it is not clear if the comparisons provide sufficient information to enhance understanding of the underlying issues that generate the observed values of water productivity.

Recalling the framework presented above, we can depict observations of water productivity in coordinate space defined either by crop yields and applied water, or by crop yields and ET, or water consumed. Choosing the first of these, suppose we observe three estimates of water productivity, perhaps in three locations or on three farms. We can represent these observations as points A, B, and C in Figure 6. Recalling that the slope of a ray from the origin, through each point, depicts the water productivity for each, we can readily observe that the water productivity for observation A is greater than that for observation B, which is greater than that for observation C. A strict preference for points reflecting higher water productivity would result in the choice of Point A as most desirable. Yet that choice might be inappropriate. Observing the points, alone, does not provide information regarding the underlying production functions. It is possible that any of the points represents an inefficient use of water, or that all points represent efficient use. We need to know more about the production setting to determine which point or points might be desirable.

Suppose the underlying production functions are those depicted in Figure 7. It would appear that the production situation at Point A is least desirable, as Point A lies to the right of the point of maximum yield. Thus, the same amount of yield could be generated with less applied water. Point B might also be inefficient, as it appears production is very close to the point of maximum yield. If there is a positive opportunity cost of the applied water, it is likely that the incremental cost exceeds the incremental again at Point B. It is therefore possible that Point C is the most desirable point of those depicted in Figure 7, even though the estimated water productivity at that point is the smallest.

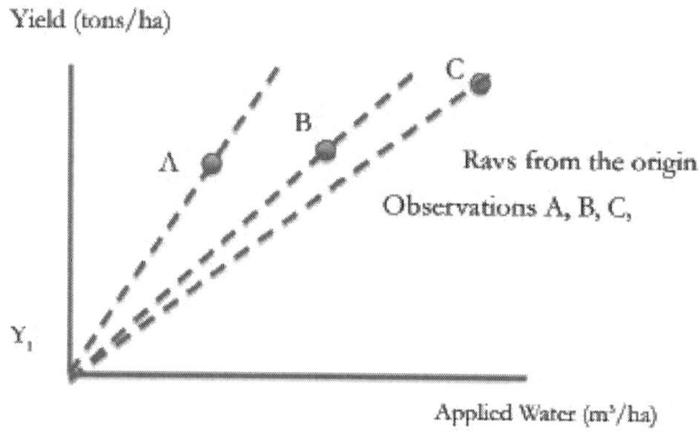

Figure 6. Comparing observations of water productivity, with respect to applied water.

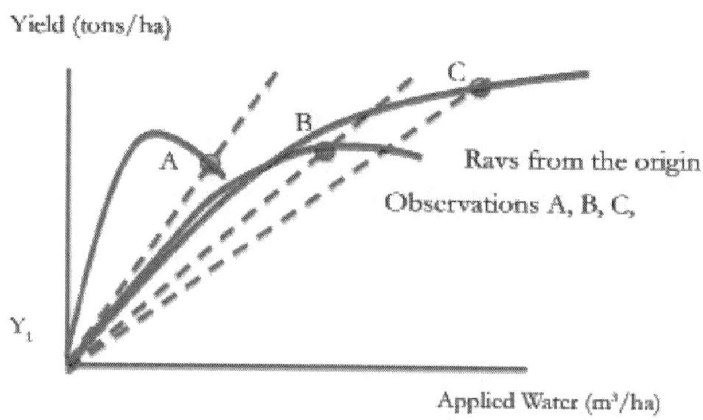

Figure 7. Considering the underlying production functions.

This is just one of many possible depictions of the underlying production functions that might generate the observations represented by Points A, B, and C. Differences in climate, soils, irrigation systems, and human capital can generate an array of empirical production functions within a given region. Observed values of crop yields might lie on the same function, or they might all lie on different functions, as depicted in Figure 7. In any event, it is not possible to evaluate the desirability of Points A, B, and C using only the measure of water productivity. This perspective can be demonstrated also in yield-ET space.

Suppose we observe three production situations, depicted as Points A, B, and C in Figure 8. We can see by inspection that Point A has the highest measure of water productivity, followed by Point B, and then Point C. Yet we could not conclude that Point A is preferable, without knowing the incremental costs and benefits of moving from Point A to Point B. If the incremental cost of providing the additional ET is less than the incremental gain, Point B would be preferred to Point A. The same analysis pertains to the potential move from Point B to Point C. The information provided by the estimated water productivity is not sufficient for determining which of these points is best from either the farm-level or societal perspective.

It is quite possible also that we observe situations in which farmers are operating on different production functions, due to differences in the amount or quality of the other, non-water inputs available to them. For example, farmers with better access to high-quality seeds, fertilizer, and pesticides will have the ability to generate higher yields per unit of transpired water. Their production functions will lie above those pertaining to farmers with limited access or with less financial ability to purchase and apply complementary inputs in a timely fashion. This information, while pertinent to discussions of agricultural policy and livelihoods, will not be reflected in estimates of water productivity.

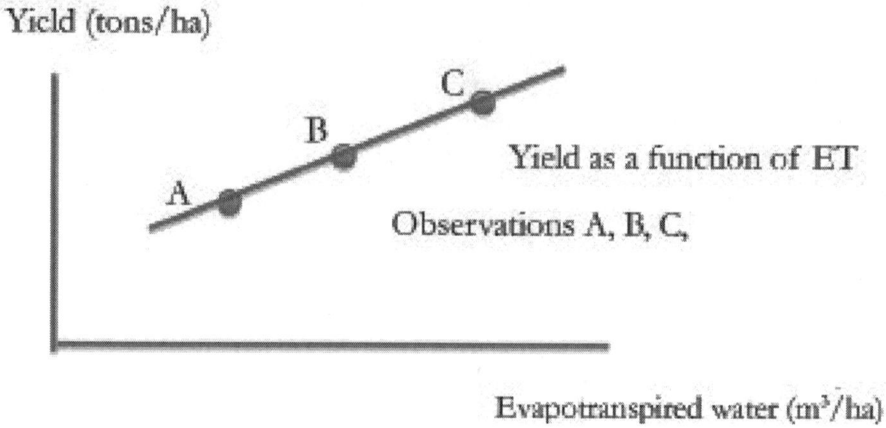

Figure 8. Comparing observations of water productivity, with respect to ET.

We might observe points such as those depicted in Figure 8, yet they might lie along different production functions, as depicted in Figure 9. The policy challenges implied by these depictions are different. In the case of Figure 8, the challenge is to determine whether or not additional transpiration is desirable, and then how to motivate greater use of water per hectare. The challenge implied by Figure 9 is that of determining how to motivate farm-level shifts in the yield-ET function. Investments in efforts to provide farmers with more affordable access to complementary inputs might be among the pertinent policy actions.

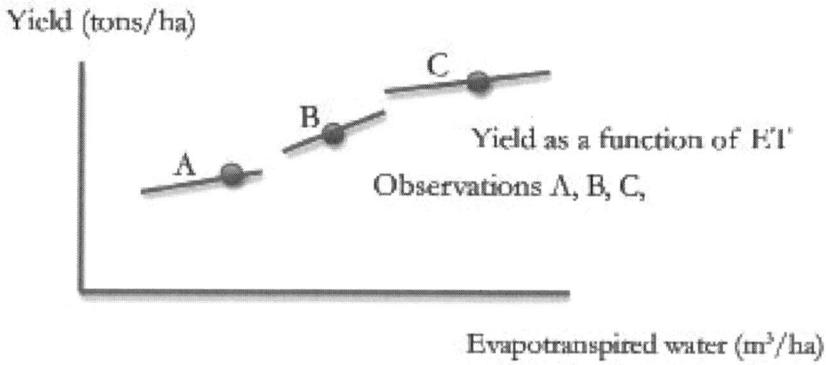

Figure 9. Comparing observations of water productivity, with respect to ET.

The situation depicted in Figure 9 suggests also that estimates of water productivity from different points in time must be interpreted with care. Over time, with improvements in agricultural technology and human capital, we expect to see crop water production functions shift upward. As this occurs, it is likely that transpiration rates per hectare also will increase, coincident with higher crop yields. Observations such as Point C in Figure 9 likely will replace observations such as Points A and B. As depicted inFigure 9, the higher output per hectare comes at the cost of greater evapotranspiration per hectare, such that the measure of water productivity is smaller. Yet the value of the additional output might exceed the opportunity cost of the additional water evapotranspired. In the future, we might aspire to observe yield-ET combinations that lie above and to the left of Point B, if advances in plant genetics enable the production of higher yields with less evapotranspiration.

An Empirical Example

Li and Barker 41 describe the higher crop yields and improvements in water management made possible by the alternate wetting and drying method of rice irrigation (AWD), in which the soil is allowed to dry between irrigation events. The authors suggest that water deliveries and losses to percolation and seepage are reduced when using the alternate wetting and drying method. They provide estimates of crop water productivity for traditional and AWD irrigation in four provinces of China. Those data are helpful in depicting the difficulty in gaining insight when comparing estimates of crop water productivity over time, locations, and crop types.

The data in Table 4 were compiled originally by authors other than Li and Barker [41]. Thus, it is not possible to know the numbers of observations that generate each value in Table 4 or to know the characteristics of the underlying distributions. Yet, taken on face value, the data reflect several interesting comparisons, within and across provinces, and over time. For example, in Hubei, the yield of traditional, middle-season rice in 1998 was higher than the yield of traditional, early-season rice in 1997, and yet the estimated water productivity is higher in 1997. Similarly, in Zhejiang, the yield of traditional, late-season rice is higher than the yield of traditional, early-season rice in 1998, yet the estimated water productivity is higher for the early-season rice. These are certainly plausible observations. The question is whether or not the estimates of water productivity are helpful in determining which production setting is better, or which should be preferred over the other.

Table 4. Rice yields and estimates of farm-level water productivity for rice, irrigated with traditional methods and using alternate wetting and drying. Source: This table is reproduced from the paper by Li and Barker (2004) 41

Province	Year	Rice Type	Yield (kg/ha)		CWP (kg/m^3)	
		(season)	**Traditional**	**AWD**	**Traditional**	**AWD**
Hubei	1997	Early	5,438	6,393	0.889	1.183
	1998	Middle	7,767	8,790	0.856	1.122
Guangxi	1980	Early	6,975	7,518	1.333	1.688
	1980	Early	6,876	7,242	1.217	1.519
Shandong	1986	Middle	8,799	9,954	1.006	1.419
	1989	Middle	8,242	10,359	0.853	1.858
Zhejiang	1998	Early	6,075	6,746	1.168	1.775
	1998	Late	6,363	6,746	0.994	1.599

Looking across provinces, the yields of traditional rice are higher in Shandong than in Guangxi, yet the estimates of water productivity are higher for Guangxi. Lacking information regarding the opportunity cost of water in the two provinces and the costs of other, complementary inputs, it is not possible to determine if the higher values of water productivity in Guangxi are meaningful from a policy perspective. Similarly, the yields of AWD rice in Zhejiang are smaller than in Guangxi, yet the estimates of water productivity are higher for Zhejiang.

The yields and water productivities for AWD rice generally are higher than the yields and productivities for traditionally irrigated rice (Table 4). Yet without information describing the incremental costs and benefits associated with the change in water management, it is not possible to know if the higher yields and water productivities are desirable. For example, if weeding costs are higher or more pesticides are needed when using alternate wetting and drying, the incremental cost of achieving higher yields might negate the value of the incremental gain. In any event, it is not possible to derive helpful conclusions from the limited information presented in Table 4. Much more is needed, regarding the opportunity costs of key inputs and the stochastic nature of crop yields, rainfall, percolation, ET, and the resulting estimates of water productivity.

SUMMING UP

The motivation for increasing water productivity is well-stated and compelling. With increasing competition for freshwater resources, farmers, researchers, and public officials must continuously pursue opportunities to improve agricultural productivity, while promoting also the wise use of renewable and non-renewable resources. Estimates of water productivity seem pertinent, particularly in water scarce areas. It is indeed compelling to think about increasing the returns per unit of water consumed in agriculture. Yet estimates of water productivity, by themselves, do not provide the information needed to fully understand the production opportunities, constraints, and livelihood considerations that determine farm input choices and the resulting probabilistic distributions of crop and livestock output.

Many authors compare estimates of water productivity across locations and over time. Yet the estimates are point observations, calculated by dividing some measure of crop or livestock production by some measure of water applied or consumed. It is not possible to derive helpful insight from such comparisons, without knowing much more about the production settings, the availability of complementary inputs, the distribution of rainfall, and the risk and uncertainty farmers face when making decisions regarding crop choices, input levels, and markets. We need also to know more about the incremental costs and gains of water use in each setting, rather than simply comparing measures of average productivity.

In summary, estimates of water productivity do not enhance understanding of farm-level water management in rainfed and irrigated settings in a manner that would guide researchers or policy makers toward wise interventions and investments. Water interacts with many other inputs in crop and livestock production. Thus, to ensure sufficient food production in future we must analyze production settings in greater detail, measuring interactions involving water and other inputs, and evaluating strategies that acknowledge those interactions, just as farmers must do each season. We must also invest greater effort in understanding risk and uncertainty, particularly from the perspective of smallholder farmers, who have little freeboard for accommodating crop failures or recovering from mistakes in input allocation. We can develop more useful guidance for smallholders if we enhance our analysis in ways that look beyond point estimates of water productivity.

ACKNOWLEDGMENTS

I appreciate the helpful comments of two reviewers, who encouraged me to enhance the clarity of discussion in several sections of the paper.

REFERENCES

1. Rockström, J.; Barron, J. Water productivity in rainfed systems: Overview of challenges and analysis of opportunities in water scarcity prone savannahs. *Irrig. Sci.* **2007**, *25*, 299–311.
2. Molden, D.; Oweis, T.; Steduto, P.; Bindraban, P.; Hanjra, M.A.; Kijne, J. Improving agricultural water productivity: Between optimism and caution. *Agric. Water Manag.* **2010**, *97*, 528–553.
3. Cai, X.; Molden, D.; Mainuddin, M.; Sharma, B.; Ahmad, M.-U.-D.; Karimi, P. Producing more food with less water in a changing world: Assessment of water productivity in 10 major river basins. *Water Int.* **2011**, *36*, 42–62.
4. Jia, S.; Ge, Z.; Fang, X. Less water, more grain in dry Hebei Province, China. *Water Int.* **2011**, *36*, 861–872.
5. Brauman, K.A.; Siebert, S.; Foley, J.A. Improvements in crop water productivity increase water sustainability and food security—A global analysis. *Environ. Res. Lett.* **2013**, *8*.
6. Kijne, J.; Barker, R.; Molden, D. *Water Productivity in Agriculture: Limits and Opportunities for Improvement*; CABI Publishing: Wallingford, UK, 2003.
7. Liu, J.; Williams, J.R.; Zehnder, A.J.B.; Yang, H. GEPIC—Modelling wheat yield and crop water productivity with high resolution on a global scale. *Agric. Syst.* **2007**, *94*, 478–493.
8. Vince, G. Getting more drops to the crops. *Science* **2010**, *327*, 800.

9. Monaghan, J.M.; Daccache, A.; Vickers, L.H.; Hess, T.M.; Weatherhead, E.K.; Grove, I.G.; Knox, J.W. More 'crop per drop': Constraints and opportunities for precision irrigation in European agriculture. *J. Sci. Food Agric.* **2013**, *93*, 977–980.
10. Borlaug, N.E. Feeding a hungry world. *Science* **2007**, *318*, 359.
11. De Fraiture, C.; Wichelns, D. Satisfying future water demands for agriculture. *Agric. Water Manag.* **2010**, *97*, 502–511.
12. Fedoroff, N.V.; Battisti, D.S.; Beachy, R.N.; Cooper, P.J.M.; Fischhoff, D.A.; Hodges, C.N.; Knauf, V.C.; Lobell, D.; Mazur, B.J.; Molden, D.; *et al.* Radically rethinking agriculture for the 21st century. *Science* **2010**, *327*, 833–834.
13. Turral, H.; Svendsen, M.; Faures, J.M. Investing in irrigation: Reviewing the past and looking to the future. *Agric. Water Manag.* **2010**, *97*, 551–560.
14. You, L.; Ringler, C.; Wood-Sichra, U.; Robertson, R.; Wood, S.; Zhu, T.; Nelson, G.; Guo, Z.; Sun, Y. What is the irrigation potential for Africa? A combined biophysical and socioeconomic approach. *Food Policy* **2011**, *36*, 770–782.
15. Wichelns, D. Investing in small, private irrigation to increase production and enhance livelihoods. *Agric. Water Manag.* **2014**, *131*, 163–166.
16. Passioura, J. Increasing crop productivity when water is scarce: From breeding to field management. *Agric. Water Manag.* **2006**, *80*, 176–196.
17. Zoebl, D. Is water productivity a useful concept in agricultural water management? *Agric. Water Manag.* **2006**, *84*, 265–273.
18. Zwart, S.J.; Bastiaanssen, W.G.M.; de Fraiture, C.; Molden, D.J. A global benchmark map of water productivity for rainfed and irrigated wheat. *Agric. Water Manag.* **2010**, *97*, 1617–1627.
19. Rockström, J.; Karlberg, L.; Wani, S.P.; Barron, J.; Hatibu, N.; Oweis, T.; Bruggeman, A.; Farahani, J.; Qiang, J. Managing water in rainfed agriculture: The need for a paradigm shift. *Agric. Water Manag.* **2010**, *97*, 543–550.
20. Araya, A.; Stroosnijder, L. Assessing drought risk and irrigation need in northern Ethiopia. *Agric. For. Meteorol.* **2011**, *151*, 425–436.

21. Burney, J.A.; Naylor, R.L. Smallholder irrigation as a poverty alleviation tool in sub-Saharan Africa. *World Dev.* **2012**, *40*, 110–123.
22. Burney, J.A.; Naylor, R.L.; Postel, S.L. The case for distributed irrigation as a development priority in sub-Saharan Africa. *Proc. Natl. Acad. Sci. USA* **2013**, *110*, 12512–12517.
23. De Fraiture, C.; Giordano, M. Small private irrigation: A thriving but overlooked sector. *Agric. Water Manag.* **2014**, *131*, 167–174.
24. Zwart, S.J.; Bastiaanssen, W.G.M. Review of measured crop water productivity values for irrigated wheat, rice, cotton and maize. *Agric. Water Manag.* **2004**, *69*, 115–133.
25. Igbadun, H.E.; Ramalan, A.A.; Oiganji, E. Effects of regulated deficit irrigation and mulch on yield, water use and crop water productivity of onion in Samaru, Nigeria. *Agric. Water Manag.* **2012**, *109*, 162–169.
26. Zhang, H.; Oweis, T. Water-yield relations and optimal irrigation scheduling of wheat in the Mediterranean region. *Agric. Water Manag.* **1999**, *38*, 195–211.
27. Zhang, H.; Wang, X.; You, M.; Liu, C. Water-yield relations and water-use efficiency of winter wheat in the North China Plain. *Irrig. Sci.* **1999**, *19*, 37–45.
28. Zand-Parsa, S.; Sepaskhah, A.R. Optimal applied water and nitrogen for corn. *Agric. Water Manag.* **2001**, *52*, 73–85.
29. Tolk, J.A.; Howell, T.A. Field water supply: Yield relationships of grain sorghum grown in three USA Southern Great Plains soils. *Agric. Water Manag.* **2008**, *95*, 1303–1313.
30. Steduto, P.; Hsiao, T.C.; Raes, D.; Fereres, E. AquaCrop—The FAO crop model to simulate yield response to water: I. Concepts and underlying principles. *Agron. J.* **2009**, *101*, 426–437.
31. Karam, F.; Kabalan, R.; Breidi, J.; Rouphael, Y.; Oweis, T. Yield and water-production functions of two durum wheat cultivars grown under different irrigation and nitrogen regimes. *Agric. Water Manag.* **2009**, *96*, 603–615.
32. Sadras, V.; Roget, D.; Krause, M. Dynamic cropping strategies for risk management in dry-land farming systems. *Agric. Syst.* **2003**, *76*, 929–948.

33. Blum, A. Drought resistance, water-use efficiency, and yield potential: Are they compatible, dissonant, or mutually exclusive? *Aust. J. Agric. Res.* **2005**, *56*, 1159–1168.
34. Kirkegaard, J.A.; Hunt, J.R. Increasing productivity by matching farming system management and genotype in water-limited environments. *J. Exp. Bot.* **2010**, *61*, 4129–4143.
35. Biazin, B.; Sterk, G.; Melesse Temesgen, M.; Abdulkedir, A.; Stroosnijder, L. Rainwater harvesting and management in rainfed agricultural systems in sub-Saharan Africa: A review. *Phys. Chem. Earth* **2012**, *47–48*, 139–151.
36. Torkamani, J.; Shajari, S. Adoption of new irrigation technology under production risk. *Water Resour. Manag.* **2008**, *22*, 229–237.
37. Takeshima, H.; Yamauchi, F. Risks and farmers' investment in productive assets in Nigeria. *Agric. Econ.* **2012**, *43*, 143–153.
38. Rockström, J.; Barron, J.; Fox, P. Rainwater management for increased productivity among small-holder farmers in drought prone environments. *Phys. Chem. Earth* **2002**, *27*, 949–959.
39. Hussain, I.; Sakthivadivel, R.; Amarasinghe, U. Land and water productivity of wheat in the western Indo-Gangetic Plains of India and Pakistan: A comparative analysis. In *Water Productivity in Agriculture: Limits and Opportunities for Improvement*; Kijne, J., Barker, R., Molden, D., Eds.; CABI Publishing: Wallingford, UK, 2003.
40. Yan, N.; Wu, B. Integrated spatial-temporal analysis of crop water productivity of winter wheat in Hai Basin. *Agric. Water Manag.* **2014**, *133*, 24–33.
41. Li, Y.; Barker, R. Increasing water productivity for paddy irrigation in China. *Paddy Water Environ.* **2004**, *2*, 187–193.

Chapter 8

Sustainable Water Management in Urban, Agricultural, and Natural Systems

Tess Russo [1,2], Katherine Alfredo [1] and Joshua Fisher [3]

[1]Columbia Water Center, Columbia University, 500 W 120th Street, New York, NY 10027, USA

[2]Department of Geosciences, The Pennsylvania State University, 310 Deike, University Park, PA 16802, USA

[3]The Advanced Consortium on Cooperation, Conflict, and Complexity (AC4), Columbia University, 475 Riverside Dr. #253, New York, NY 10115, USA

ABSTRACT

Sustainable water management (SWM) requires allocating between competing water sector demands, and balancing the financial and social resources required to support necessary water systems. The objective of this review is to assess SWM in three sectors: urban, agricultural, and natural systems. This review explores the following questions: (1) How is SWM defined and evaluated? (2) What are the challenges associated with sustainable development in each sector? (3) What are the areas of greatest potential improvement in urban and agricultural water management systems? And (4) What role does country development status have in SWM practices? The methods for evaluating water management practices range from relatively simple indicator methods to integration of multiple models, depending on the complexity of the problem and resources of the investigators. The two key findings and recommendations for meeting SWM objectives are: (1) all forms of water must be considered usable, and reusable, water resources; and (2) increasing agricultural crop water production represents the largest opportunity for reducing total water consumption, and will be required to meet global food security needs. The level of regional development should not dictate sustainability objectives, however local infrastructure conditions and financial capabilities should inform the details of water system design and evaluation.

INTRODUCTION

Water is at the foundation of sustainable development as it is the common denominator of all global challenges: energy, food, health, peace and security, and poverty eradication.—UN Water [1]

Definitions

Sustainable development is commonly defined by the Brundtland Report [2] as meeting the needs of the present generation without compromising the ability of future generations to meet their own needs. "Needs" include economic, environmental and ecosystem service delivery, and cultural goals including identity and subjectively defined values. Together, these are commonly referred to as the sustainability triple bottom line. Sustainable development is the combination of sustaining the natural environment, resources, and community, and development of the economy and societal goals.

Sustainable water management (SWM) is a critical component of sustainable development, and accounts for similar issues as sustainability. Mays [3] defines SWM as meeting current water demand for all water users without impairing future supply. More specifically, SWM should contribute to the objectives of society and maintain ecological, environmental, and hydrologic integrity [4]. The definition proposed by Alley et al. [5] for groundwater management cites protection of the components in the sustainability triple bottom line: environment, economy, and society. A more holistic objective is provided in Agenda 21 [6] which ensures that "adequate supplies of water of good quality are maintained for the entire population of the planet, while preserving the hydrological, biological and chemical functions of ecosystems, adapting human activities within the capacity limits of nature and to combat vectors of water-related diseases."

Definitions of SWM present challenges for adoption because of the widespread use of subjective language and lack of detail. Most definitions offer only a broad conceptualization of the sectors or environments to consider. The language in most SWM definitions is usually qualitative and often normative. For example, sustainability entails "the design of human and industrial systems to ensure that humankind's use of natural resources and cycles does not lead to diminished quality of life due either to losses in future economic opportunities, or to adverse impacts on social conditions, human health, and the environment" [7]. This definition uses "diminished quality of…" and "adverse impacts on…" to describe thresholds of sustainable system effects, which are difficult to interpret for management purposes. With respect to sustainable urban water management in Agenda 21 from the United Nations, Larsen & Gujer [8] assert that no concept proposed is applicable in practice. The normative guidance of most sustainability definitions is especially problematic for policy makers and water managers who are motivated to adopt sustainable practices, but have little tangible support.

Integration of Sustainable Development and Sustainable Water Management

Sustainable development and SWM are inherently related due to the requirement of water for development [9,10]. Water is a fundamental requirement for human life and well-being, thus proper management of water is a means to improve food production, reduce poverty and water-related diseases. SWM involves allocating water between competing purposes and users. This allocation can be represented as a hierarchy, similar to Maslow's hierarchy of human needs (Table 1) [11]. The framework asserts that the foundational biophysical needs must be met before effort can be dedicated to fulfilling higher needs. The second level of Maslow's hierarchy is "safety", which translates to security of domestic water and agricultural production at the local level. The third and fourth levels of needs in this hierarchy, "social" and "esteem", correspond to broader community water services and a focus on maintenance, justice, and responsibility, respectively. The two hierarchies share commonalities from levels 1 to 4, and seemingly end at the fifth level. Maslow's inward-looking "self-fulfillment" contrasts with resource sustainability, which emphasizes an outward-looking perspective including the fulfillment of other users both now and in the future. The greatest challenge for developing countries is promoting SWM, the fifth and final need, while still working to address their foundational needs.

Table 1. Comparison of Maslow's hierarchy of human needs and the hierarchy of water management needs. Adapted from Melloul and Collin [11]

Level	Maslow's Hierarchy of Human Needs	Hierarchy of Water Management Needs
5	Self-fulfillment	Water resource sustainability
4	Esteem (status, recognition)	National water projects (supply, remediation, public awareness)
3	Social (family, community)	Regional water projects (supply, treatment plants)
2	Safety (security, stability, law, order)	Local development (agriculture, domestic water, water quality standards)

1	Physiological for survival (Air, water, food, shelter, procreation)	Biophysical individual needs (water for survival)

This review explores urban, agricultural, and natural water system management in developing and developed nations. We aim to address the following questions: (1) How is SWM defined and evaluated? (2) What are the challenges associated with sustainable development in each sector? (3) What are the areas of greatest potential improvement in urban and agricultural systems? And (4) What role does country development status have in SWM? The first question is addressed generally in Section 1 and Section 2, and more specifically with examples in Section 3. Questions (2)–(4) are addressed in Section 3 and Section 4, and are summarized in Table 2, Table 3 and Table 4. Limitations to this review are discussed in Section 4.3. Rather than providing exhaustive discussion on all aspects of water management, the objective of this study is to conceptualize the dialogue around what constitutes SWM practices according to three sectors: urban, agricultural, and natural systems. Municipal and agricultural water use account for over 80% of global human water use, and higher in many developing nations [12]. The number of people experiencing perennial water shortages in urban areas is expected to increase from 150 million (ca. 2000) to 993 million by 2050, and those experiencing seasonal water shortages will rise from 1 to 3.1 billion, respectively [13]. Agricultural water use accounts for 70% of global water consumption, and can exceed 90% in developing nations. Water management of natural systems is a key requirement of sustainable urban and agricultural management practices. These three sectors represent the majority of water use globally, but may not constitute the majority of water use in a specific region or country (e.g., Canada, Russia, and the United Kingdom). Therefore, a more detailed study of SWM should be completed at the regional scale to provide the most relevant information for management and policy purposes.

Table 2. Primary sustainable water management objectives

Sector	Developing Areas	Developed Areas
Urban	Equitable delivery Reliability System flexibility with growth	Supporting demand Infrastructure longevity Recycling and reuse Environmental protection
Agriculture	Food security Expansion of irrigated area Supplemental irrigation Crop water productivity	Crop water productivity Environmental protection Resource conservation
Ecosystem	Protection of valued ecosystem services	Protection and/or restoration of natural functions within development constraints

Table 3. Developing country challenges and solutions for sustainable water management in urban, agricultural, and natural systems

Sector	Challenges	Examples +
Urban	Intermittent operation Lost or stolen water Rapid urban growth Political conflict	India [14], Ethiopia [15] Palestine [16] China [17] *Mid East* [16,18,19]
Agriculture	Irrigation infrastructure cost Subsidies promoting irresponsible use Low water use efficiency	*Sub-Saharan Africa* [20] India [21] *N Africa* [22]; China, Pakistan [23]
Ecosystem	Economic development priorities	India [24]
Sector	**Solutions**	**Examples**

Urban	Institutional improvements	Algeria [25]; Palestine [26]
	Low-tech water capture and treatment *	Greece [27]
	Graywater reuse *	Jordan [28]
	Cooperation, sharing riparian rights	*Mid East* [29]; *World* [30]
	Stakeholder engagement	Mexico [31]; *World* [32]
Agriculture	Optimize water productivity *	*World* [33,34]
	Improve subsidy/pricing structure	India [35]; *Mid East* [36]; *World* [37]
	Supplemental irrigation with rainwater harvesting	*Sub-Saharan Africa* [38,39]; Burkina Faso [40]; *Drylands* [41]
Ecosystem	Communication of ecosystem service value	*Africa* [42,43]; Ethiopia [44]

Notes: + Regions and non-country area descriptions are in *italics*; * Topics covered in the discussion.

Table 4. Developed country challenges and solutions for sustainable water management in urban, agricultural, and natural systems

Sector	**Challenges**	**Examples** +
Urban	Demand management	Australia [45]; USA [46]
	Cost of adopting new technology	*World* [47]
Agriculture	Over-allocation of water resources	S Africa [48]
	Cost of new technology *vs.* low cost of water	USA [49,50]
Ecosystem	Development priorities	*World* [51]
Sector	**Solutions**	**Examples**
Urban	Conservation education	S Africa [52]; Australia [53]
	Water reuse and recycling *	*World* [54]; UK [55]
	Stormwater management, green infrastructure *	UK [56]; Denmark [57]
Agriculture	Improve water allocation, irrigation management	Greece [58]; Spain [59]
	Affordable precision agriculture methods *	USA [49]

Ecosystem	Identification and protection of most valuable resources Definition of a degradation tolerance level	USA [60] *World* [61]

Notes: + Regions and non-country area descriptions are in *italics*; * Topics covered in the discussion.

EVALUATION TOOLS AND MODELS

Sustainable systems are especially apt to compare favorably with conventional systems when the comparison includes a full cost accounting of the environmental and public health harms and benefits of each system.—Horrigan et al. [62]

SWM evaluation requires accounting for real costs, opportunity costs, and competing requirements among and within water use sectors. These sometimes include vague political and socio-economic components, which often do not translate easily to the quantitative values necessary for planning, decision-making, and rigorous monitoring and assessment. Evaluation models range from generic indices used to compare management practices across multiple agencies and environments, to site-specific modeling analyses that enable individual managers and governments to assess progress toward or away from sustainable practices [63].

Inputs to the evaluation should be quantifiable, independent from each other, unambiguous, and representative. Quantitative evaluations allow for assessing conditions and comparing management options. Following the definitions for sustainable development, many tools aim to evaluate environmental, social, and economic aspects of the system [64,65]. While many such tools quantify environmental impact and resource utilization, we continue to lack robust quantitative evaluation methods for social-cultural criteria, and the interconnected impacts of social, biological, and physical components of complex systems [66,67].

In addition to having a quantitative framework that bridges a complex system, evaluation models must account for the time and place dependent conditions of sustainability. Time is relevant within the lifecycle of a specific action or management practice, and also in the subsequent impacts on the environment. The spatial scale of evaluation governs what resources are assumed to be available for a given location, ranging from household to village to watershed to country to multinational region. The spatial extents of the region, including quantity of available water resources and competing users, will influence analysis of existing and potential development. Accounting for local conditions and perspectives in the evaluation method has particular utility when assessing policy impacts over time or when advocating for specific policy changes [25].

There are three primary evaluation methods for SWM: (1) indicators and indices; (2) product related assessment; and (3) integrated assessment [64]. Water indicators and indices should simplify, quantify, and communicate information [68]. The classification, weighting, and method of index aggregation varies by model [68], making the processes somewhat arbitrary. However, the advantage of the indicator method is a simple numerical result that provides comparative capability across cases. Common indices which address water in sustainable development include the Water Poverty Index [69], the Canadian Water Sustainability Index [70], the Environmental Performance Index [71], and the Watershed Sustainability Index [72]. Water management indicators typically include water infrastructure, environmental quality, economics and finance, institutions and society, human health, and technology [4]. Other indices which specifically address some of the socio-economic dilemmas mentioned previously include accessibility of data, institutional schemes to resolve water conflict, and democratic water-related decision making [3].

Product related assessments, or life cycle assessments (LCA), can provide information about land, water, and energy requirements for a physical system or supply chain. The LCA framework can be used to inventory a set of sustainability indicators across the supply chain of a water system. Examples such as the Ecological Footprint [73] and its hydrologic corollary, the Water Footprint [74], translate biophysical assets into progress measurements. These tools are resource accounting mechanisms that enable integrated quantitative assessments of land and water resources in terms of current and potential demand, and regenerative probability at national and global scales.

Integrated assessments tend to be holistic assessments completed using system dynamics models, risk analysis, cost-benefit analysis, and impact assessments. Integrated assessments provide a systems perspective, often incorporating more robust quantification than the indicator method alone. Information theory can be used to study how efficiently human systems are using resources, and the resilience of those systems [75]. A systems dynamics modeling approach can be used with a framework based on viability loops to monitor water system acceptance, use, and economics [76]. Chung and Lee [77] demonstrate the value of coupling a hydrologic model and a multi-criteria decision model (MCDM) to evaluate alternatives for sustainable development. Many of the integrated assessment methods can include a Monte Carlo method to test for uncertainty and sensitivity.

Beyond the three typical evaluation types, groundwater use is often evaluated by comparing extraction to the aquifer safe yield. Historically, groundwater extraction sustainability was based only on groundwater recharge, which over-simplified subsurface dynamics. Early definitions of safe yield preclude pumping that is "dangerous" [78] or produces an "undesirable result" [79], including rapid declines in groundwater levels. More recent studies have called into question both the value and the sustainability of safe yield [80,81,82]. Evaluations of groundwater development sustainability account for natural groundwater recharge rates, as well as capture which includes induced recharge and decreases in natural discharge [5,83,84]. Advances in numerical and statistical models are improving our estimates and projections of groundwater use sustainability.

The diversity of evaluation methods described above provides water users and managers a wide array of tools to assess the sustainability of water use and allocation in specific water systems at a variety of scales. These tools reflect the complexity of evaluating sustainability within the physical and social world. Local variations, data availability, and socio-political objectives may lead to selection of different water quantity sustainability evaluation methods. This is particularly relevant for setting estimates of thresholds for sustainable use and allocation. Future work will improve evaluation frameworks in terms of measuring and assessing SWM including spatial and temporal efficiency, supply longevity, and equitable distribution.

SECTOR REVIEWS

Developing Countries: Practices, Challenges, and Solutions

Urban Water Development

Access to potable water is among the most important prerequisites for healthy and productive development. Water and socioeconomic development are integrally connected—either creating a vicious downward spiral or reinforcing positive development practices [85]. Sustainable development in urban areas requires reliable, equitable, and easily accessible water (Table 2). Providing water to the rapidly growing urban populations in developing nations creates a complex logistic and economic problem. This is exacerbated in areas where urban expansion is largely informal an unplanned, as demand is difficult to project and consumption is difficult to monitor.

Indicator methods are commonly used for evaluating urban water management. Developing successful indicator methods requires continued efforts to quantify the relationships between urban water management and environmental sustainability, which groups including the UN Centre for Human Settlements and the UN Environmental Program are currently pursuing. Monitoring and evaluation of urban water systems in developing regions will help improve current conditions and inform future development. One assessment evaluated 25 indicators in four criteria groups: technical, social/environmental, financial, and institutional [86]. They found that technical (design flow and the functionality of the system) and social criteria (status of use, equity, decision making in operations and maintenance) are more crucial to sustaining a water system than financial and institutional criteria, although all are necessary.

Urban water system sustainability faces challenges emerging from physical resource, infrastructure, and socioeconomic conditions (Table 3). In water rich regions, water distribution systems are optimized to meet a specified minimum pressure, whereas in water stressed regions equal water distribution among users is the goal. In the latter case, distribution systems should be designed with the expectation of intermittent operation rather than assuming continuous supply, thus reducing pressure problems and inequities in distribution [14]. Both water rich and poor regions are faced with the infrastructure design and water supply challenges of meeting demands associated with population growth [17]. Infrastructure and operation problems include leaking pipes and "stolen" water in water scarce regions. Material and construction failures, electricity supply or operation and maintenance issues all affect water distribution. Thirty percent of water supply systems serving rural sub-Saharan communities are non-functioning at any given time [15]. Challenges include infrastructure issues, however their origins may include socio-institutional problems involving responsibility, knowledge, and coordination [87].

Regional conflict presents a significant challenge to SWM, especially in countries with differing development levels. Melloul and Collin [11] give an example where two factions of a local population are at different stages of the Water Management Hierarchy of Needs (Table 1). These disparities led to conflicts concerning resource allocation and development plans. Political conflict often leads to increased water stress, and vice versa. The UN has numerous programs involved in assessment and management of water resources, including transboundary integrated water resources management and urban water supply [88]. Recent evaluations of cooperation and competition in transboundary water management show a clear preference for cooperation rather than conflict among riparian neighbors [29,30].

Increasingly, sustainability solutions include participatory planning and water capture and reuse. Sustaining a water system requires personnel training, monitoring, health and safety regulations, and communication with users [25]. Sustainable water utilities should involve the population which they serve, and utilize the local knowledge of water resources and traditional management practices [89]. The inclusion of women in water management decisions is particularly important in cultural contexts where women are traditionally charged with collecting water for domestic consumption and use, as water management can have immediate and long-term impacts on gender equality [85,90].

Water capture, storage, and reuse are becoming common aspects of SWM systems [91]. While traditional capture and treatment systems may be cost-prohibitive for many developing regions, there are low-cost, low-technology treatment systems including constructed wetlands or layering with indigenous rock materials which can be effective for treating water for reuse [27,92,93]. Local conditions and capacity for operation require evaluation to determine the feasibility on a site by site basis [94]. Reuse of graywater in urban areas is becoming more common in water stressed, developing regions [28]. Decentralized systems are recommended at the household or neighborhood scale for producing recycled water for gardening or domestic reuse.

Agricultural Water Development

Food production is an integral part of sustainable human development. Sustainable agricultural water management objectives include attaining food security and maximizing food water productivity in rainfed and irrigated agriculture (Table 2). A majority of global population growth is occurring in developing countries, some of which rely heavily on rainfed agriculture. Therefore, rainfed agriculture must expand and/or see increased productivity to feed the growing population [95]. Irrigation provides a water supply buffer during dry periods, reducing the risks to purely rainfed agriculture. Though irrigation led to notable production gains in India and China over the past decades, it was coupled with increased dependency on unrenewable water resources in many regions.

Quantification of agricultural practice impact on water resources will help prevent these negative impacts under future development. Pereira *et al.* [34] present case studies for Tunisia and China using an indicator method for evaluating agricultural management strategies. Their framework accounts for water reuse and crop production economics. Dixon and Wood [44] used optimization of social, environmental, and economic benefits to define agricultural plans. While leaving natural areas uncultivated was environmentally sustainable, it provided little support for human development. The optimization method recommended mixed use: leaving part of the land in its natural condition, and converting part of it for cultivation. Despite the demonstrated success, optimizing agricultural development and environmental water resources is not typical.

Few of the Green Revolution nations prioritized sustainable agricultural water use with increasing yields and meeting food security goals. Therefore, while the Green Revolution more than doubled crop yields, resulting in greater food security, lower food prices, and improved social and economic conditions [96], it also had many unintended environmental consequences. Government subsidies for water and energy (for pumping water) have led to wider adoption of irrigation and higher crop yields, however the low costs provide little incentive for farmers to conserve water and adopt more efficient irrigation methods. In many agricultural regions, greater than 100% of the renewable water resources are used for irrigation each year [97]. These regions must assess alterative irrigation support policies and agricultural water management strategies in order to sustain production in the future [21].

Solutions include improving water productivity and reducing risk in rainfed crops by using supplemental irrigation (Table 3). Postel [33] suggests a transition from Green Revolution to Blue Revolution, where productivity is measured in crop production per unit water. In much of the developing world, irrigation systems are less than 50% efficient. Increasing irrigation productivity requires improvements to technical, managerial, institutional, and agronomic methods. In rainfed agricultural regions, improvements in rainwater capture, selection of drought-resistant crops, and alternate tillage practices will be critical to feeding growing populations [22]. Using rainwater harvesting for supplemental irrigation to mitigate the impact of dry spells, combined with soil fertility management, can more than double rainfed crop yields [38,40]. Though the practice is not broadly used, there is widespread potential for water harvesting for supplemental irrigation in many rainfed agricultural regions [39,98].

Environmental Protection

Evaluation of environmental sustainability is required concurrently with development planning to protect ecosystem services (Table 2). Integrated modeling is a common approach for evaluating impacts between the human and natural environments. For example, integrated modeling can be used to assess the impact of rainwater harvesting and storage on downstream hydrology and environmental conditions

[99]. Application of this model can reduce conflicts between upstream and downstream water stakeholders, while maintaining environmental flow requirements. A case study in northern India demonstrates a progressive water resource development plan to evaluate conservation of land, water, and forest resources [24]. Their proposed plan considers ecological sustainability, socio-economic options and user requirements. An important and uncommon component of their proposed plan is ensuring income generation where land use change will affect employment options. This is critical given the emphasis on land conservation and restoration in an area where 77% of the population depends on agriculture for income, similar to much of the rural developing world.

An effective method to encourage conservation and land protection is communicating the economic value of natural resources [42,44], thus allowing a community to make the optimal development decision (Table 3). Communities that depend heavily on the natural environment for their livelihoods innately place a high priority on sustainable use of resources and protection of the environment [43]. Quantifying ecosystem value makes it simple to determine whether a use scenario protecting water resources is most beneficial for stakeholders.

Relatively few studies explore sustainable human development and water in the natural environment in developing countries, with the exceptions of the examples above, the Millennium Ecosystem Assessment report [10], and the bi-annual release of the Environmental Performance Index [71]. The priority for many developing regions is meeting present basic needs, which, according to the Water Management Hierarchy (Table 1), must be addressed prior to planning for a sustainable future. Generally, protection of the environment is not integrated into policy as it is considered anti-development. To reduce the risks of this policy omission, the Southern African Development Community has adopted the fundamental assumption that, "strategies to reduce poverty should not lead to further degradation of water resources or ecological functions and services." They hope to move mainstream sustainability criteria into water resources policy and management, especially in developing nations.

DEVELOPED COUNTRIES: PRACTICES, CHALLENGES, AND SOLUTIONS

Urban Water Development

Every human being has a right to clean water. For urban areas, our vision is water management where water and its constituents can be safely used, reused and returned to nature.—Hellström et al. [67]

The primary functions of urban water management include meeting hygiene (supply and sanitation), drinking water, drainage, urban agriculture, and recreational needs. Meeting these needs while protecting natural resources and human health, especially if water is scarce, is key to sustainable development [100]. Challenges include water demand management, and the cost of adopting sustainable, innovative technologies (Table 4).

Indicator methods were most common for evaluating sustainable urban water management, similar to the developing country studies. The Sustainable Urban Water Management Program in Sweden uses five categories of sustainability criteria. The criteria include: nontoxic environment, health and hygiene, human resources, natural resources, financial resources, functionally robust and flexible, adaptable to local conditions, and easy to understand to encourage responsible user behavior [67]. The structure of the evaluation framework is modular so it can be used to evaluate cities, structures, and scenarios. Several cities have developed their own sustainability evaluation models, including the Sustainable Index for Taipei and Sustainable Seattle, with 51 and 40 indicators, respectively. Unfortunately, development often leads to some level of environmental or resource degradation. "Sustainable" practices in these cases may be defined as having the least detrimental environmental, economic, and social impacts. For example, to account for continued development in the water-scarce western United States, proposed water policy would use a hierarchical tree of environmental values to direct damage to the least valuable resource [60].

In developed countries where water demand far exceeds sustenance requirements, water demand management becomes relevant. The need for conservation and demand management is a function of resource availability, population growth, and climate change. In addition to technology innovation and pricing incentives, education and awareness are critical to achieving conservation goals. One case study concluded that 50% of time and resources need to be spent on user education and awareness [52]. Marlowet *al.* [47] review sustainable urban water management methods, and discuss probable barriers including the uncertainty and challenges of innovative technology adoption, financial considerations, and institutional bias or advocacy. Innovative systems may require experimentation and carry inherent risk, so as more cities adopt sustainable practices, sharing experiences and results is required to broaden adoption. The cost of infrastructure and technology improvements may be offset by water prices which reflect its true value.

Future solutions for sustainable urban water management include broader adoption of water reuse and green infrastructure practices. Approaches for improving sustainable water reuse can be categorized into two groups: (1) Substitution (e.g., rainwater capture and reuse of graywater); and (2) regeneration (e.g., potable reuse and wastewater reuse) [54]. Public awareness of treatment practices and cooperation for how treated water can safely be used are key to developing successful reuse programs [53]. In addition to treating and reusing wastewater, stormwater runoff should be viewed as a potential water source, rather than water that needs to be treated and routed away [91]. Ellis [56] advocates for a wider green infrastructure framework including green roofs, downspout disconnection, and canopy cover of 25%–30% in riparian corridors. These broader sustainability practices can improve urban water management, while also reducing energy consumption, and benefiting local ecosystem services.

Agricultural Water Development

Agricultural water use accounts for 70% of worldwide consumption. Irrigation is broadly adopted in developed regions because it supports higher crop yields with lower risk [101]. Similar to developing regions where irrigation is occurring, many aspects of industrial agriculture in the first world are unsustainable [62]. Evaluation of water management in agricultural development must account for inevitable negative impacts on the environment [80,102]. Evaluating the sustainability of irrigation practices requires determining the degree of environmental degradation that people are willing to accept. Metrics for evaluating irrigation systems and on-farm water use practices must account for off-farm (or downstream) consequences. SWAGMAN is a tool that integrates hydrologic, agricultural, environmental, and economic models [103]. The SWAGMAN model helps determine crop types and irrigation methods which increase agricultural value while improving environmental conditions relative to business as usual. Sustainable irrigation practices will always be site specific, depending on local climate, soil, topography, and water source. For example, areas with shallow water tables will need different strategies compared to areas with deep water tables. For the former, water logging and soil salinization may be primary concerns, while energy costs and groundwater depletion are issues for the latter.

Negative environmental impacts due to agriculture are inevitable, but they can be minimized by increasing the efficiency of nutrient additions and water use [104]. There are clear opportunities for improving water productivity in the agricultural sector. Two primary areas include: (1) improving water allocation; and (2) improving application efficiency (Table 4). Improving allocation requires selection of crops which require less irrigation per unit of production, given the local climate and growing conditions. For example, reducing beef production, which requires 100 times the water required to produce equal protein from grains [105], would improve water productivity. Broad improvements in crop water productivity requires a better understanding of the biological, biogeochemical, and ecological processes active in agricultural settings. Improved farm management practices include no-till farming, drip or other efficient irrigation technologies, and soil tensiometers or moisture sensors. Many farms in developed nations can afford to use irrigation, though only ~10% of farmers in the Western U.S. improve their irrigation application timing by using soil monitoring technology which informs irrigation scheduling based on actual conditions and plant requirements [49]. The low adoption rate of precision agricultural technologies is evidence of the cost and implementation barriers faced by farmers. However, as aquifers become depleted, the cost of pumping water from deep aquifers will increase which may make conservation technologies relatively affordable. The solution requires efficient irrigation technologies which are economically preferable to current water use costs.

Environmental Protection

Water management must balance development needs with environmental protection or restoration (Table 2). Multiple methods for ecosystem or environmental evaluation have been developed in both applied and academic literatures (e.g., [105,106,107]). Typical water-related evaluation criteria include stream discharge, groundwater elevation, lake stage, and rates of flow between surface, soil, and groundwater. Some of these criteria have single thresholds for defining sustainable hydrologic behavior, while others have more complicated dynamic requirements. Numerous models can be used for estimating the sustainability of riverine systems in developed watersheds. The framework Ecological Limits of Hydrologic Alteration (ELOHA) is based on the requirement of natural flow variation, *i.e.*, that a single minimum flow threshold is insufficient [108]. The quantity of flow in the river, or the flow regime, can be characterized using look-up tables with hydrologic modeling [109]. In an effort to recreate natural flow variability on a controlled river, statistical methods can be used to compare current flow frequency distributions to a natural reference [110].

Several integrated models aim to identify natural water system vulnerability and value under human development scenarios to help prioritize protection and management (Table 4). Given the natural or desired streamflow regime, the impact of population growth and development can be modeled and used to inform water management practices [111]. When applied over multiple watersheds, results can identify locations where water and growth management strategies are most needed, and where they will provide the greatest benefit. Alternatively, modeling can identify incompatibilities between human and ecosystem requirements, which are then used to develop and test new management strategies [107]. The timing and frequency of incompatibilities must be assessed using both within-year and among multiple year analyses to account for seasonal and long term trends in water supply and demand.

Given that urban and agricultural development will impact environmental systems, defining sustainability requires setting a tolerance level or acceptable pathway. Instead of setting a static goal, Chapin *et al.* [61] propose a method of ecosystem stewardship that maintains a trajectory which sustains social-ecological systems and ecosystem services. The trajectory may vary over time as conditions change, and actions should be taken to correct undesirable trajectories when they occur. The temporal variability of this method makes it perhaps more difficult to define, however the flexibility is valuable for accommodating population growth and climate change.

DISCUSSION

Areas of Greatest Improvement in Urban and Agricultural Systems

Urban Systems: All Water is a (Re) Usable Resource

Meeting the challenges of water resources sustainability increasingly involves ... applying innovative approaches to conjunctive use of groundwater, surface water, artificial recharge, and water reuse.—Alley and Leake [80]

Managing water resource sustainability requires considering water in all states and forms as potential resources for use and reuse. Improving use efficiency, capture, and reuse of these non-traditional water resources is more critical in water stressed regions, and those which are expected to become stressed due to climate change or population growth. Treating wastewater is a key part of solving water scarcity [16]. As climate change makes dry regions drier [112,113], the need for water capture and reuse intensifies in areas with increasing water stress.

There are two key points when considering all water as resources; first, not all applications require the same quality water, and second, not all "used" water requires the same level of treatment before it can be reused. Treatment before and treatment after of the combined water stream uses unnecessary amounts of energy and effort [91].

Incentivizing a selective system of treatment and reuse requires that water be priced appropriately. Water must be considered an economic good to account for its competing uses [85], where the price depends on availability and quality. To encourage treatment and reuse, the value of water should equal the cost of treating source water to necessary standards.

Green infrastructure and stormwater capture are not fully utilized in both developing and developed regions. Excess water during rain events can be harvested and stored for use in dry periods [91]. Depending on the level of existing infrastructure and water application purposes, the scale of water capture and distribution can range from household to neighborhood to city. In many developing regions, small scale capture is recommended at the household level using storage tanks or infiltration ponds in conjunction with hand pumps for recovery [114]. Increased capture and use of all available water resources will significantly reduce water stress, especially during dry seasons and periods of drought.

Agricultural Systems: Crop Water Productivity

Improvements in crop water productivity can result in commensurately large decreases in water use because agriculture accounts for the largest quantity of water use. With water use efficiency ranging between 10% and 30% for rainfed and between 40% and 95% for irrigated agriculture [22,115,116], there is nearly always opportunity for improvement. Methods for improved on-farm agricultural water management include supplementing rainfed crops, irrigation scheduling, and efficient irrigation methods [117]. At the national or global scale, agricultural water use efficiency can be improved by growing more food in high water productivity regions and exporting to less productive regions [118]. Irrigation (or electricity) subsidies should target regions with sustainable water sources, or should couple incentives for high efficiency irrigation systems and low water-requirement crops. As water demands increase in the developing world, irrigation reliability is expect to decline from 0.79 (out of 1.0) in 2005 to 0.71 in 2025 [119]. In areas where groundwater use is unsustainable, improving efficiency (and decreasing total extraction) allows production to continue longer into the future [120].

The use of technology to inform irrigation scheduling can save water, and also increase crop yields compared to over-irrigation [121]. The estimated benefits of irrigation scheduling will vary by method and location. Methods for irrigation scheduling include using soil moisture sensors and incorporating weather forecast data. Soil moisture sensors or tensiometers indicate soil wetness conditions, and can be compared to plant moisture requirements. Sensor informed agriculture water savings range from 18% to 50% [122]. Crop water production can be further improved by combining irrigation scheduling with farm management techniques including mulching, reducing soil hydrophobicity, and the use of wastewater [123]. Instrument cost is a primary barrier in both developing and developed regions. The technology must be designed within the budget of the intended farmers, and/or should be subsidized by the government.

In regions where water is the limiting factor to production instead of land, increases in water use efficiency may allow farmers to irrigate more land. While this is arguably not a SWM solution because it does not reduce the total amount of water used for irrigation [124], it does increase the crop water productivity of the region. In cases were over-irrigation results in runoff and water supply for downstream users, irrigation reduction may in fact reduce water availability to these users, and should be considered in the overall management strategy [125].

Relevance of Country Development Status

This review highlights several differences as well as similarities between developing and developed nation SWM objectives (Table 2), challenges, and solutions (Table 3 and Table 4). The differences lie in the context and level of development, and not in the definition of sustainability. The model selected for SWM evaluation typically varies to accommodate local infrastructure and economic conditions, but still maintains the objectives of sustainable development.

For urban water systems, equitable and reliable supply is the objective in developed and developing regions. Water stress and aging infrastructure are challenges faced around the world. Developing nations may face additional challenges including intermittent electricity and disparities in access to water delivery or built infrastructure. Indeed, the focus of SWM in developing countries is on providing equitable and reliable water supply, while developed nations may focus on water reuse and system longevity (Table 2). In addition, many developed regions now strive to have water systems that mimic natural ecosystems [75]. Water system evaluation may also include a demand management component in developed regions. Conversely, this metric is irrelevant in regions where people are not yet receiving the recommended amount of water, for example more than half of the population of the Middle East, as determined by the Islamic Network on Water Resources Development and Management [16].

Evaluation models and opportunities in agriculture are most sensitive to whether the farm is rainfed or irrigated, and what technology and information the farmer has access to. In developing nations, increasing crop productivity and equitable water allocation is critical. However, as we have learned from the Green Revolution in South Asia, increasing productivity at the expense of natural resources is not a sustainable solution. In rainfed agriculture, methods for improving yields should include supplemental irrigation during critical growth periods, and on-farm management practices that improve soil moisture holding capacity. In developed nations with high levels of food security, increasing resource use efficiency should be a priority. The use of soil and plant sensors to inform irrigation scheduling must be part of the solution, as well as switching to more efficient irrigation technologies.

The literature on environmental water management is dominated by developed region case studies and models where water management needs have been met and sustainability emerges as a priority. We note the presence of environmental objectives in both urban and agricultural development (Table 2). Conversely, in developing regions, the objectives for environmental water management are presented by two contrasting sides. The first follows the order of the Hierarchy of Water Management Needs (Table 1), asserting that until individual and community water needs are met, environmental sustainability is not a priority. Larsen and Gujer [8] stated "sustainable development is only possible in the absence of extreme poverty ... In areas with a lack of safe drinking water, biological diversity and other ecosystem requirements will not be given any priority." In alignment with this thinking, meaningful protection of the environment is generally not integrated into developing nation policy because it is considered anti-development.

It is irresponsible to allow damage to resources that will ultimately be required for a population to continue developing. The second perspective, which is held by communities who depend heavily on the natural environment for their livelihoods, prioritizes sustainable resource use and protection of the environment regardless of economic development level. The Southern African Development Community is an example of a progressive group regarding environmental sustainability. They maintain that poverty reduction does not need to compromise environmental health and services. This perspective will grow as further evidence of economic and social development coupled with environmental protection is successfully documented. All regions can have the intent of SWM, while practices may vary with geography and economic capabilities.

Limitations

Limitations of this study include: (1) the exclusive scope of urban, agricultural, and natural systems; (2) the challenge of obtaining municipal and other types of non-peer reviewed documents; and (3) omission of relevance of country geography, climate, and other factors. While urban and agricultural systems account for over 80% of global water consumption, other water uses such as industrial and recreational can be significant in some countries or local regions. Future reviews of other water uses would be complementary to this one, and provide value in improving water management practices across sectors. Much of the relevant literature, especially for urban water management, may include white papers, municipal reports, and other documents which are not typically available in academic databases. This review likely overlooked numerous documents describing SWM assessments, practices, challenges, and solutions. Future reviews would benefit from a more thorough search for these reports. Lastly, the scope of this paper did not include a synthesis of how country geography, climate, and other factors affect SWM practices. Future studies may focus on a number of factors to illustrate why SWM practices can differ in proximal nations or be similar in distant nations; relevant endogenous factors may include in-country distribution of wealth and resources, socio-cultural traditions, and political stability, while exogenous factors may include climate, geology, and a more detailed look at historical and present inter-country conflict.

CONCLUSIONS

SWM of urban, agricultural, and environmental systems is integral to continued development. Numerous models and metrics exist for evaluating sustainable management practices. Improvements to these methods should focus on the interconnectedness of social and physical systems using robust *quantitative* metrics. Urban water management in developing regions faces challenges of equitable delivery, especially under rapid urban population growth. Sustainable management plans should focus on continued improvements in stakeholder involvement and infrastructure in developing regions, and on water reclamation and reuse in developed regions. Water reuse will reduce stress during drought periods, though technology adoption cost and risks are still barriers in both developing and developed nations. Improvements to crop water productivity can benefit all sectors of water users discussed in this article by reducing competition between the agricultural sector and urban and environmental users. Crop water production in irrigated areas can be improved with changes in crop water allocation and adoption of efficient irrigation and on-farm technologies, while rainfed agricultural areas will benefit from supplemental irrigation. Maintaining sustainable water supply in natural systems can be seen to conflict with development practices if only looking at the near-term future. Long-term economic development is clearly linked to environmental system health, evidenced by developed country focus on restoration and protection of water resources. In application, decisions informed by the estimated value of ecosystem services may be used to set thresholds for environmental degradation, in the context of social and economic development goals. SWM will vary with geography and economic capabilities, though all regions can manage water resources in a way that supports sustainable social, economic, and environmental development.

ACKNOWLEDGMENTS

This work was supported by the Advanced Consortium on Cooperation, Conflict and Complexity, The Columbia Water Center, and the Earth Institute Fellowship program at Columbia University. We thank two anonymous reviewers for their constructive comments, and Madhura Karnik and Kuan-Lin Liu for assembling the preliminary collection of literature.

REFERENCES

1. Rio+20 from a Water Perspective. Available online: http://www.unwater.org/news-events/news-details/en/c/207616/ (accessed on 1 February 2014).
2. United Nations World Commission on Environment and Development (UN WCED). Our Common Future: Report of the World Commission on Environment and Development; Document A/42/427. United Nations: New York, NY, USA, 1987.
3. Mays, L. Water Resources Sustainability; McGraw-Hill Professional: New York, NY, USA, 2006.
4. Loucks, D.; Gladwell, J. Sustainability Criteria for Water Resource Systems; Cambridge University Press: Cambridge, UK, 1999.
5. Alley, W.; Reilly, T.; Franke, O. Sustainability of Ground–Water Resources; US Department of the Interior, US Geological Survey: Denver, CO, USA, 1999.
6. United Nations. Agenda 21. In Proceedings of the United Nations Conference on Environment & Development, Rio de Janero, Brazil, 3–14 June 1992.
7. Mihelcic, J.R.; Crittenden, J.C.; Small, M.J.; Shonnard, D.R.; Hokanson, D.R.; Zhang, Q.; Chen, H.; Sorby, S.A.; James, V.U.; Sutherland, J.W.; et al. Sustainability science and engineering: The emergence of a new metadiscipline. Environ. Sci. Technol. 2003, 37, 5314–5324.
8. Larsen, T.; Gujer, W. The concept of sustainable urban water management. Water Sci. Technol. 1997, 35, 3–10. CrossRef

9. Gleick, P. Basic water requirements for human activities: Meeting basic needs. Water Int. 1996, 21, 83–92. CrossRef
10. Ecosystems and Human Well-Being: Wetlands and Water Synthesis; Millennium Ecosystem Assessment: Washington, DC, USA, 2005.
11. Melloul, A.J.; Collin, M.L. Harmonizing water management and social needs: A necessary condition for sustainable development. The case of Israel's coastal aquifer. J. Environ. Manag. 2003, 67, 385–394.
12. FAO AQUASTAT. Water withdrawal by sector, around 2006. Available online: www.fao.org/nr/water/aquastat/globalmaps/AquastatWorldDataEng_20121214_Withdrawal.pdf (accessed on 1 February 2014).
13. McDonald, R.I.; Green, P.; Balk, D.; Fekete, B.M.; Revenga, C.; Todd, M.; Montgomery, M. Urban growth, climate change, and freshwater availability. Proc. Natl. Acad. Sci. USA 2011, 108, 6312–6317.
14. Vairavamoorthy, K.; Akinpelu, E. Design of sustainable water distribution systems in developing countries. ASCE World Water Congr. 2001, 44, 1–10.
15. Rahmato, D. Water Resource Development in Ethiopia: Issues of Sustainability and Participation; Forum for Social Studies: Addis Ababa, Ethiopia, 1999.
16. Al-Qurashi, A.; Husain, T. Sustainable Water Resources Development Plan for the Middle-East Countries. In Proceedings of the ASCE World Water Congress, Orlando, FL, USA, 20–24 May 2001.
17. Jiang, Y. China's water scarcity. J. Environ. Manag. 2009, 90, 3185–3196.
18. Gleick, P. Water, drought, climate change, and conflict in Syria. Weather Clim. Soc. 2014, 6, 331–340.
19. Conde, G. An indicator of conflict? Water in between Turkey, Syria and Iraq. Reg. Cohes. 2014, 4, 81–100.
20. Inocencio, A.; Kikuchi, M.; Tonosaki, M.; Maruyama, A.; Merrey, D.; Sally, H.; de Jong, I. Costs and Performance of Irrigation Projects: A Comparison of Sub-Saharan Africa and Other

Developing Regions; International Water Management Institute: Colombo, Sri Lanka, 2007.

21. Russo, T.; Devineni, N.; Lall, U. Assessment of Agricultural Water Management in Punjab, India Using Bayesian Methods. In Sustainability of Integrated Water Resources Management: Water Governance, Climate and Ecohydrology; Setegn, S., Donoso, M., Eds.; Springer: Berlin, Germany, 2014.
22. Araus, J. The problems of sustainable water use in the Mediterranean and research requirements for agriculture. Ann. Appl. Biol. 2004, 144, 259–272.
23. Khan, S.; Tariq, R.; Yuanlai, C.; Blackwell, J. Can irrigation be sustainable? Agric. Water Manag. 2006, 80, 87–99.
24. Tiwari, P.C.; Joshi, B. Environmental changes and sustainable development of water resources in the Himalayan headwaters of India. Water Resour. Manag. 2011, 26, 883–907.
25. Benzerra, A.; Cherrared, M.; Chocat, B.; Cherqui, F.; Zekiouk, T. Decision support for sustainable urban drainage system management: A case study of Jijel, Algeria. J. Environ. Manag. 2012, 101, 46–53.
26. Daibes, F. Towards sustainable development in the water sector: A perspective from Palestine. Water Sci. Technol. 2000, 42, 81–86.
27. Antoniadis, A.; Takavakoglou, V.; Zalidis, G.; Darakas, E.; Poulios, I. Municipal wastewater treatment by sequential combination of photocatalytic oxidation with constructed wetlands. Catal. Today 2010, 151, 114–118.
28. Al-Jayyousi, O. Greywater reuse: Towards sustainable water management. Desalination 2003, 156, 181–192.
29. Voza, D.; Vukovic, M.; Carlson, L.; Djordjevi , D.B. International water conflict and cooperation: The role of power relations among. Int. J. Humanit. Soc. Sci. 2012, 2, 56–66.
30. De Stefano, L.; Edwards, P.; de Silva, L.; Wolf, A.T. Tracking cooperation and conflict in international basins: Historic and recent trends. Water Policy 2010, 12, 871–884.
31. Starkl, M.; Brunner, N.; Lo, E.; Martı, L. A planning-oriented sustainability assessment framework for peri-urban water management in developing countries. Water Res. 2013, 47, 1–9.

32. Pearson, L.J.; Coggan, A.; Proctor, W.; Smith, T.F. A sustainable decision support framework for urban water management. Water Resour. Manag. 2009, 24, 363–376.

33. Postel, S. Pillar of Sand: Can the Irrigation Miracle Last?; W.W. Norton & Company: New York, NY, USA, 1999.

34. Pereira, L.S.; Cordery, I.; Iacovides, I. Improved indicators of water use performance and productivity for sustainable water conservation and saving. Agric. Water Manag. 2012, 108, 39–51.

35. Kumar, M.D.; Scott, C.A.; Singh, O.P. Can India raise agricultural productivity while reducing groundwater and energy use? Int. J. Water Resour. Dev. 2013, 29, 557–573.

36. Dinar, A.; Mody, J. Irrigation water management policies: Allocation and pricing principles and implementation experience. Nat. Resour. Forum 2004, 28, 112–122.

37. Johansson, R.C.; Tsur, Y.; Roe, T.L.; Doukkali, R.; Dinar, A. Pricing irrigation water: A review of theory and practice. Water Policy 2002, 4, 173–199.

38. Rockström, J.; Barron, J. Water productivity in rainfed systems: Overview of challenges and analysis of opportunities in water scarcity prone savannahs. Irrig. Sci. 2007, 25, 299–311.

39. Biazin, B.; Sterk, G.; Temesgen, M.; Abdulkedir, A.; Stroosnijder, L. Rainwater harvesting and management in rainfed agricultural systems in sub-Saharan Africa—A review. Phys. Chem. Earth Parts A/B/C 2012, 47–48, 139–151.

40. Fox, P.; Rockström, J. Water-harvesting for supplementary irrigation of cereal crops to overcome intra-seasonal dry-spells in the Sahel. Phys. Chem. Earth 2000, 25, 289–296.

41. Oweis, T.; Hachum, A. Supplemental Irrigation: A Highly Efficiency Water Use Practice; International Center for Agricultural Research in the Dry Areas (ICARDA): Aleppo, Syria, 2012.

42. Schuyt, K.D. Economic consequences of wetland degradation for local populations in Africa. Ecol. Econ. 2005, 53, 177–190.

43. Hirji, R.; Mackay, H.; Maro, P. Defining and Mainstreaming Environmental Sustainability in Water Resources Management in Southern Africa—A Summary; SADC, IUCN, SARDC, World Bank: Washington, DC, USA, 2002.

44. Dixon, A.B.; Wood, A.P. Wetland cultivation and hydrological management in eastern Africa: Matching community and hydrological needs through sustainable wetland use. Nat. Resour. Forum 2003, 27, 117–129.
45. White, S.; Fane, S. Designing cost effective Water Demand Management Programs in Australia. Water Sci. Technol. 2002, 46, 225–232.
46. Dawadi, S.; Ahmad, S. Evaluating the impact of demand-side management on water resources under changing climatic conditions and increasing population. J. Environ. Manag. 2013, 114, 261–275.
47. Marlow, D.R.; Moglia, M.; Cook, S.; Beale, D.J. Towards sustainable urban water management: A critical reassessment. Water Res. 2013, 47, 7150–7161.
48. Le Maitre, D.; Colvin, C.; Maherry, A. Water resources in the Klein Karoo: The challenge of sustainable development in a water-scarce area. S. Afr. J. Sci. 2009, 105, 39–48.
49. Schaible, G.D.; Aillery, M.P. Water Conservation in Irrigated Agriculture: Trends and Challenges in the Face of Emerging Demands; US Department of Agriculture, Economic Research Service: Washington, DC, USA, 2012.
50. Ward, F.A.; Michelsen, A.M.; DeMouche, L. Barriers to water conservation in the Rio Grande Basin1. J. Am. Water Resour. Assoc. 2007, 43, 237–253.
51. Costanza, R.; D'Arge, R.; de Groot, R.; Farber, S. The value of the world's ecosystem services and natural capital. Nature 1997, 387, 253–260.
52. Buckle, J.S. Water demand management—Philosophy or implementation? Water Sci. Technol. 2000, 4, 25–32.
53. Marks, J.; Zadoroznyj, M. Managing sustainable urban water reuse: structural context and cultures of trust. Soc. Nat. Resour. 2005, 18, 557–572.
54. Grant, S.B.; Saphores, J.-D.; Feldman, D.L.; Hamilton, A.J.; Fletcher, T.D.; Cook, P.L.M.; Stewardson, M.; Sanders, B.F.; Levin, L.A.; Ambrose, R.F.; et al. Taking the "waste" out of "wastewater" for human water security and ecosystem sustainability. Science 2012, 337, 681–686.

55. Dixon, A.; Butler, D.; Fewkes, A. Water saving potential of domestic water reuse systems using greywater and rainwater combination. Water Sci. Technol. 1999, 39, 25–32.
56. Ellis, J.B. Sustainable surface water management and green infrastructure in UK urban catchment planning. J. Environ. Plan. Manag. 2013, 56, 24–41.
57. Pizzol, M.; Scotti, M.; Thomsen, M. Network Analysis as a tool for assessing environmental sustainability: Applying the ecosystem perspective to a Danish water management system. J. Environ. Manag. 2013, 118, 21–31.
58. Georgiou, P.E.; Papamichail, D.M. Optimization model of an irrigation reservoir for water allocation and crop planning under various weather conditions. Irrig. Sci. 2008, 26, 487–504.
59. Playán, E.; Mateos, L. Modernization and optimization of irrigation systems to increase water productivity. Agric. Water Manag. 2006, 80, 100–116.
60. Rajagopal-Durbin, A.; Durbin, T.J. Wells are not always water follies: Sustainable groundwater policies for the American west. Water Policy 2008, 10, 145–164.
61. Chapin, F.S.; Carpenter, S.R.; Kofinas, G.P.; Folke, C.; Abel, N.; Clark, W.C.; Olsson, P.; Smith, D.M.S.; Walker, B.; Young, O.R.; et al. Ecosystem stewardship: Sustainability strategies for a rapidly changing planet. Trends Ecol. Evol. 2010, 25, 241–249.
62. Horrigan, L.; Lawrence, R.; Walker, P. How sustainable agriculture can address the environmental and human health harms of industrial agriculture. Environ. Health Perspect. 2002, 110, 445–456.
63. Gleick, P.; Chaleki, E.; Wong, A. Measuring Water Well-Being: Water Indicators and Indices. In The World's Water 2002–2003: The Biennial Report on Freshwater Resources; Island Press: Washington, DC, USA, 2002.
64. Ness, B.; Urbel-Piirsalu, E.; Anderberg, S.; Olsson, L. Categorising tools for sustainability assessment. Ecol. Econ. 2007, 60, 498–508.
65. Singh, R.K.; Murty, H.R.; Gupta, S.K.; Dikshit, A.K. An overview of sustainability assessment methodologies. Ecol. Indic. 2009, 9, 189–212.

66. Mori, K.; Christodoulou, A. Review of sustainability indices and indicators: Towards a new city sustainability index (CSI). Environ. Impact Assess. Rev. 2012, 32, 94–106.
67. Hellström, D.; Jeppsson, U.; Kärrman, E. A framework for systems analysis of sustainable urban water management. Environ. Impact Assess. 2000, 20, 311–321.
68. Juwana, I.; Muttil, N.; Perera, B.J.C. Indicator-based water sustainability assessment—A review. Sci. Total Environ. 2012, 438, 357–371.
69. Lawrence, P.R.; Meigh, J.; Sullivan, C. The Water Poverty Index: An International Comparison; Keele University: Staffordshire, UK, 2002.
70. The Canadian Water Resources Sustainability Index (CWSI); Policy Research Initiative: Ottawa, ON, Canada, 2007.
71. Hsu, A.; Emerson, J.; Levy, M.; de Sherbinin, A.; Johnson, L.; Malik, O.; Jaiteh, M. Environmental Performance Index; Yale Center for Environmental Law and Policy: New Haven, CT, USA, 2014.
72. Chaves, H.M.L.; Alipaz, S. An integrated indicator based on basin hydrology, environment, life, and policy: The watershed sustainability index. Water Resour. Manag. 2006, 21, 883–895.
73. Ewing, B.; Reed, A.; Galli, A.; Kitzes, J.; Wackernagel, M. Calculation Methodology for the National Footprint Accounts, 2010 Edition; Global Footprint Network: Oakland, CA, USA, 2010.
74. Hoekstra, A.; Chapagain, A.; Aldaya, M.; Mekonnen, M. The Water Footprint Assessment Manual: Setting the Global Standard; Earthscan: London, UK, 2011.
75. Ulanowicz, R.E.; Goerner, S.J.; Lietaer, B.; Gomez, R. Quantifying sustainability: Resilience, efficiency and the return of information theory. Ecol. Complex. 2009, 6, 27–36.
76. Bagheri, A.; Hjorth, P. A framework for process indicators to monitor for sustainable development: Practice to an urban water system. Environ. Dev. Sustain. 2006, 9, 143–161.
77. Chung, E.-S.; Lee, K.S. Prioritization of water management for sustainability using hydrologic simulation model and multicriteria

decision making techniques. J. Environ. Manag. 2009, 90, 1502–1511.

78. Lee, C. The determination of safe yield of underground reservoirs of the closed basin type. Trans. Am. Soc. Civ. Eng. 1915, 78, 148–251.
79. Todd, D. Ground Water Hydrology; John Wiley: New York, NY, USA, 1959.
80. Alley, W.; Leake, S. The journey from safe yield to sustainability. Ground Water 2004, 42, 12–16.
81. Bredehoeft, J. Safe yield and the water budget myth. Ground Water 1997, 35, 929.
82. Sophocleous, M. Managing water resources systems: Why "safe yield" is not sustainable. Ground Water 1997, 35, 561.
83. Bredehoeft, J.; Durbin, T. Ground water development—The time to full capture problem. Ground Water 2009, 47, 506–14.
84. Zhou, Y. A critical review of groundwater budget myth, safe yield and sustainability. J. Hydrol. 2009, 370, 207–213.
85. Mwanza, D.D. Water for sustainable development in Africa. In The World Summit on Sustainable Development; Springer: Berlin, Germany, 2005.
86. Peter, G.; Nkambule, S.E. Factors affecting sustainability of rural water schemes in Swaziland. Phys. Chem. Earth Parts A/B/C 2012, 50–52, 196–204.
87. Brown, R.R.; Farrelly, M.A. Delivering sustainable urban water management: A review of the hurdles we face. Water Sci. Technol. 2009, 59, 839–846.
88. Töpfer, K. Balancing competing water uses—A necessity for sustainable development. Water Sci. Technol. 2003, 47, 11–16.
89. Mbilinyi, B.P.; Tumbo, S.D.; Mahoo, H.F.; Senkondo, E.M.; Hatibu, N. Indigenous knowledge as decision support tool in rainwater harvesting. Phys. Chem. Earth Parts A/B/C 2005, 30, 792–798.
90. The World's Women 2010: Trends and Statistics; United Nations Department of Economic and Social Affairs (UN DESA): New York, NY, USA, 2010.

91. Niemczynowicz, J. Urban hydrology and water management—Present and future challenges. Urban Water 1999, 1, 1–14.
92. Rahman, M.A.; Ahsan, S.; Kaneco, S.; Katsumata, H.; Suzuki, T.; Ohta, K. Wastewater treatment with multilayer media of waste and natural indigenous materials. J. Environ. Manag. 2005, 74, 107–110.
93. Angin, I. An investigation on natural wastewater treatment system and re-usability of wastewater in irrigation. J. Sustain. Agric. 2007, 31, 83–90.
94. Kivaisi, A.K. The potential for constructed wetlands for wastewater treatment and reuse in developing countries: A review. Ecol. Eng. 2001, 16, 545–560.
95. Rockstrom, J.; Barron, J.; Fox, P. Water Productivity in Rain Fed Agriculture: Challenges and Opportunities for Smallholder Farmers in Drought-Prone Tropical Agro-Ecosystems. In Water Productivity in Agriculture: Limits and Opportunities for Improvement; CABI: Wallingford, UK, 2003.
96. Falcon, W.; Naylor, R. Rethinking food security for the twenty-first century. Am. J. Agric. Econ. 2005, 87, 1113–1127.
97. Khan, S.; Hanjra, M.A. Footprints of water and energy inputs in food production—Global perspectives. Food Policy 2009, 34, 130–140.
98. Tabor, J.A. Improving crop yields in the Sahel by means of water-harvesting. J. Arid Environ. 1995, 30, 83–106.
99. Ngigi, S.N.; Savenije, H.H.G.; Gichuki, F.N. Land use changes and hydrological impacts related to up-scaling of rainwater harvesting and management in upper Ewaso Ng'iro river basin, Kenya. Land Use Policy 2007, 24, 129–140.
100. Lundin, M.; Molander, S.; Morrison, G. A set of indicators for the assessment of temporal variations in the sustainability of sanitary systems. Water Sci. Technol. 1999, 39, 235–242.
101. World Water Development Report (WWDR3) Water in a Changing World; The United Nations Educational, Scientific and Cultural Organization (UNESCO): Paris, France, 2009.
102. Wichelns, D.; Oster, J.D. Sustainable irrigation is necessary and achievable, but direct costs and environmental impacts can be substantial. Agric. Water Manag. 2006, 86, 114–127.

103. Khan, S.; Hanjra, M. Sustainable land and water management policies and practices: A pathway to environmental sustainability in large irrigation systems. L. Degrad. Dev. 2008, 487, 469–487.
104. Tilman, D.; Cassman, K.G.; Matson, P.A.; Naylor, R.; Polasky, S. Agricultural sustainability and intensive production practices. Nature 2002, 418, 671–677.
105. Pimentel, D.; Pimentel, M. Food, Energy and Society; Colorado University Press: Niwot, CO, USA, 1996.
106. Smakhtin, V.; Revenga, C.; Döll, P. A pilot global assessment of environmental water requirements and scarcity. Water Int. 2004, 29, 307–317.
107. Richter, B.D.; Mathews, R.; Harrison, D.; Wigington, R. Ecologically sustainable water management: Managing river flows for ecological integrity. Ecol. Appl. 2003, 13, 206–224.
108. Poff, N.L.; Richter, B.D.; Arthington, A.H.; Bunn, S.E.; Naiman, R.J.; Kendy, E.; Acreman, M.; Apse, C.; Bledsoe, B.P.; Freeman, M.C.; et al. The ecological limits of hydrologic alteration (ELOHA): A new framework for developing regional environmental flow standards. Freshw. Biol. 2010, 55, 147–170.
109. Acreman, M.C.; Dunbar, M.J. Defining environmental river flow requirements—A review. Hydrol. Earth Syst. Sci. 2004, 8, 861–876.
110. Arthington, A.H.; Bunn, S.E.; Poff, N.L.; Naiman, R.J. The challenge of providing environmental flow rules to sustain river ecosystems. Ecol. Appl. 2006, 16, 1311–1318.
111. Marshall, R.M.; Robles, M.D.; Majka, D.R.; Haney, J.A. Sustainable water management in the southwestern United States: Reality or rhetoric? PLoS One 2010, 5.
112. Chou, C.; Neelin, J.D.; Chen, C.-A.; Tu, J.-Y. Evaluating the "rich-get-richer" mechanism in tropical precipitation change under global warming. J. Clim. 2009, 22, 1982–2005.
113. Held, I.; Soden, B. Robust responses of the hydrological cycle to global warming. J. Clim. 2006, 19, 5686–5699.
114. Malley, Z.J.U.; Taeb, M.; Matsumoto, T.; Takeya, H. Environmental sustainability and water availability: Analyses of the scarcity and improvement opportunities in the Usangu plain, Tanzania. Phys. Chem. Earth 2009, 34, 3–13.

115. Causapé, J.; Quílez, D.; Aragüés, R. Irrigation efficiency and quality of irrigation return flows in the Ebro River Basin: An overview. Environ. Monit. Assess. 2006, 117, 451–461.
116. Falkenmark, M.; Rockström, J. The new blue and green water paradigm: Breaking new ground for water resources planning and management. J. Water Resour. Plan. Manag. 2006, 132, 129–132.
117. Scanlon, B.R.; Jolly, I.; Sophocleous, M.; Zhang, L. Global impacts of conversions from natural to agricultural ecosystems on water resources: Quantity versus quality. Water Resour. Res. 2007, 43.
118. Yang, H.; Wang, L.; Abbaspour, K.C.; Zehnder, A.J.B. Virtual water trade: An assessment of water use efficiency in the international food trade. Hydrol. Earth Syst. Sci. 2006, 10, 443–454.
119. Rosegrant, M.; Cai, X. Global water demand and supply projections global water demand and supply projections Part 2. Results and prospects to 2025. Water Int. 2002, 27, 170–182.
120. Steward, D.R.; Bruss, P.J.; Yang, X.; Staggenborg, S.A.; Welch, S.M.; Apley, M.D. Tapping unsustainable groundwater stores for agricultural production in the high plains aquifer of Kansas, projections to 2110. Proc. Natl. Acad. Sci. USA 2013, 110, E3477–E3486.
121. Irmak, S. Plant Growth and Yield as Affected by Wet Soil Conditions due to Flooding or Over-Irrigation; University of Nebraska-Lincoln Extension: Lincoln, NE, USA, 2014.
122. Van Iersel, M.; Seymour, R.M.; Chappell, M.; Watson, F.; Dove, S. Soil moisture sensor-based irrigation reduces water use and nutrient leaching in a commercial nursery. Proc. South. Nurs. Assoc. Res. Conf. 2009, 54, 17–21.
123. Stroosnijder, L.; Moore, D.; Alharbi, A.; Argaman, E.; Biazin, B.; van den Elsen, E. Improving water use efficiency in drylands. Curr. Opin. Environ. Sustain. 2012, 4, 497–506.
124. Scott, C.A. The water-energy-climate nexus: Resources and policy outlook for aquifers in Mexico. Water Resour. Res. 2011, 47.
125. Ward, F.; Pulido-Velazquez, M. Water conservation in irrigation can increase water use. Proc. Natl. Acad. Sci. USA 2008, 105, 18215–18220.

Chapter 9

Quality of Water and Antibiotic Resistance of Escherichia coli from Water Sources of Hilly Tribal Villages with and without Integrated Watershed Management—A One Year Prospective Study

Sandeep S. Nerkar[1,2] Ashok J. Tamhankar[1,2] Smita U. Khedkar[3] and Cecilia Stålsby Lundborg[1]

[1]Department of Public Health Sciences, Global Health (IHCAR), Karolinska Institutet, Stockholm, SE 17177, Sweden;

[2]Department of Environmental Medicine, Indian Initiative for Management of Antibiotic Resistance, R. D. Gardi Medical College, Ujjain, 456010, India

[3]Bac-test Laboratory, College Road, Nashik, 422005, India;

ABSTRACT

In many hilly tribal areas of the world, water scarcity is a major problem and diarrhoea is common. Poor quality of water also affects the environment. An integrated watershed management programme (IWMP) aims to increase availability of water and to improve life conditions. Globally, there is a lack of information on water contamination, occurrence of diarrhoea and antibiotic resistance, a serious global concern, in relation to IWMP in hilly tribal areas. Therefore, a prospective observational study was conducted during 2011–2012 in six villages in a hilly tribal belt of India, three with and three without implementation of an IWMP, to explore quality of water, diarrhoeal cases in the community and antibiotic resistance of Escherichia coli from water sources. The results showed that physico-chemical quality of water was within limits of safe consumption in all samples. The odds of coliform contamination in water samples was 2.3 times higher in non-watershed management villages (NWMV) compared to integrated watershed management villages (IWMV) (95% CI 0.8–6.45, $p = 0.081$). The number of diarrhoeal cases (18/663 vs. 42/639, $p < 0.05$) was lower in IWMV as compared to NWMV. Overall E. coli isolates showed high susceptibility to antibiotics. Resistance to a wider range of antibiotics was observed in NWMV.

INTRODUCTION

In many parts of the world, including India, tribal people live in hilly terrain [1]. Water scarcity is their main concern and contamination of water and diarrhoeal cases are common. Less availability and poor quality of water affects the environment as well as various life conditions. Open dug wells are most often the source of water in these tribal areas [2] and such wells are prone to physico-chemical and microbial contamination. This poses a risk to human health. Bacteria are often responsible for water contamination and subsequent disease. Antibiotics are used in the control of bacterial infections, but are also commonly unnecessarily used for non-bacterial infections.

Development of resistance by bacteria makes antibiotics ineffective for those in need of treatment [3] and this is now of serious global concern. Further, water contaminated by resistant bacteria poses the risk of spread of antibiotic resistance in the environment [4].

A watershed is an area or a region which contributes rainfall water to a river or lake. Watershed management, if implemented effectively, can help to solve the scarcity of water in that area. Classically, watershed management mainly focuses on soil and water conservation with an aim to increase availability of water [5]. The modern concept of integrated watershed management (IWM), however, besides soil and water conservation, also includes plantation of trees, improving agriculture, improving hygiene and sanitation, formation of self-help groups and proper utilization and management of available natural resources [6].

Globally, there is a lacuna of information on water contamination and antibiotic resistance in relation to watershed management, particularly in hilly tribal areas. Therefore this study aimed to explore the quality of water, occurrence of bacterial contamination in water, incidence of diarrhoeal cases, antibiotic use in the community and antibiotic resistance of Escherichia coli found in the water sources of hilly tribal villages in the context of integrated watershed management. E. coli was selected as an indicator bacterial species, as it is commonly used to study the bacterial contamination in water and also for studying antibiotic resistance patterns [7].

EXPERIMENTAL SECTION

Study Setting and Design

This prospective observational study was conducted over a period of one year from July 2011 to June 2012 in a tribal belt of Thane district of Maharashtra state, India. The rainy season occurs in this area during June to September and the remaining months is the dry season. The terrain in this area is hilly and the area is characterized by scarcity of water, except for during the rainy season. This suggests a need for conservation and proper management of water. Six tribal villages located within a radius of 15 km were chosen purposively for the study. In one set of three villages, an integrated watershed management programme (IWMP) had already been implemented. The components of the implemented IWMP were soil and water conservation, plantation of trees, improvement in agricultural and livestock production, awareness programme on hygiene and sanitation, formation of self-help groups and community participation in all developmental activities. These villages are hereafter referred to as integrated watershed management villages (IWMV). The remaining three villages, hereafter referred to as non-watershed management villages (NWMV), were located in an adjoining watershed, where no watershed management programme had been implemented.

Open dug community wells were the major source of water in both IWMV and NWMV. All together a total of seven such wells were identified and labeled. The four wells of IWMV were labeled IWMV1 to IWMV4, whereas the three wells of NWMV were labeled NWMV1 to NWMV3.

Two water samples, both of 250 mL were collected from each well at monthly interval. The samples were collected in sterilized glass bottles following standard procedures [8]. Immediately after collection, the first sample was evaluated using a field level rapid testing kit (WT023, Hi-Media, Mumbai, India) for eight physico-chemical water quality parameters namely, pH, turbidity, chlorides, total hardness, nitrates, iron, residual chlorine and fluoride [9]. The second water sample was kept in a thermo-insulated icebox (4 °C) and was transferred to the microbiology laboratory for analysis of bacterial contamination.

Enumeration of Total Coliform and Identification of Escherichia coli

For estimation of total coliform count of bacteria in water samples, the Most Probable Number (MPN) method was used. Multiple tube technique was employed [10]. In brief, Mac-Conkey broth medium was used in 15 test tubes, in which inverted Durham tubes were added for gas collection. The observation on acid and gas production in each row of five tubes was noted and coliform count was recorded using McCrady's table. E. coli was differentiated from other fecal coliforms using selective medium, eosin methylene blue agar (in this medium, the colonies of E. coli show typical metallic sheen with a dark center), followed by Eijkman test [11]. Identified E. coli isolates were further confirmed by the Vitek 2 (bioMerieux, Marcy l'Etoile, France) compact instrumentation [12].

Antibiotic Susceptibility

Confirmed E. coli isolates were tested for their antibiotic susceptibility by using the Vitek 2 instrument. Two cards, AST-GN25 and AST-EXN7 were selected, in which a total of 34 antibiotics were present for testing (Table A1). Antibiotics included were: tetracyclines, penicillins, all generation cephalosporins, carbapenems, trimethoprim/sulfamethoxazole and quinolones. Care was taken to include antibiotics that were used in the study area. The Vitek 2 instrument is an automated antibiotic susceptibility testing system that enables rapid determination of Minimum Inhibitory Concentrations (MICs) [13,14]. It uses an Advanced Expert System (AES) that provides standardized interpretative reading of MICs [15]. Updated AES parameters were set in the system to utilize default interpretation guidelines of the Clinical and Laboratory Standards Institute (CLSI) and natural resistance. Multi-Drug Resistance (MDR), defined as "resistance to three or more antibacterial classes" [16], was also evaluated.

Collection of Data on Occurrence of Diarrhoea and Antibiotic Use

The data on the occurrence of diarrhoeal cases was collected at regular intervals during the study period. Information was also collected on number of toilets from each household of the study villages once during the study period. Data on antibiotics delivered to sub-centres of primary health centres (PHCs) in the study villages from 2008–2012 was obtained from records. This data was used as proxy for antibiotics consumed by the population in the study villages and is presented as Defined Daily Doses (DDDs)/1,000 inhabitants/year. The DDD is the assumed average maintenance dose per day for a drug used for its main indication in adults [17].

Data Analysis

Data was entered in Excel version of MS-Office 2010 and analyzed using Stata 12.1 (Stata Corp., College Station, TX, USA). Descriptive statistics was used to present the data. Odds ratio (OR) was used to determine the strength of association between IWMP and coliform contamination of water. Pearson's chi-square test was used to test statistical significance as applicable. A value of $p < 0.05$ was considered statistically significant while a p-value between 0.05 and 0.10 was considered as of "borderline significance". A pilot study was conducted for one month in the study area before the main study.

Ethical Approval

Ethical approval was obtained from the institutional ethics committee of R. D. Gardi Medical College, Ujjain (No. 175/2011). Permission was also obtained from concerned local authorities for the collection of water samples and collection of data related to the study.

RESULTS AND DISCUSSION

Out of the seven wells included in the study, one well from IWMV (IWMV4) was dry for three months, from April to June and one well from NWMV (NWMV2) was dry for five months, from February to June. During this period, the villagers used other wells in their respective areas, already included in the study. In total, 76 water samples, 45 from IWMV and 31 from NWMV, were collected over a period of one year.

Table 1 shows physico-chemical quality parameters of each well of the six tribal villages. The pH of all samples was in the range of 7.0 to 7.4. Turbidity in all samples was below 5 Nephelometric Turbidity Units (NTU) except for samples from the sources IWMV2 and NWMV2, which were in the range of 5–25 and 0–10 NTU respectively during the year. Chloride content was in the range of 20–60 ppm in IWMV and 20–30 ppm in NWMV. There was not much difference in total hardness of water in both the settings and it ranged between 50–200 ppm during the whole year. Nitrate content was below 10 ppm constantly throughout the year in all samples except IWMV1 and IWMV2, where a maximum value of up to 45 ppm was recorded in the months of June and July. There were no traces of iron, residual chlorine or fluoride in any of the samples.

Table 1. Physico-chemical quality parameters (average/month) of water from each well of six tribal villages with and without implementation of integrated watershed management programm

Parameter	Integrated Watershed Management Villages (IWMV)				Non-Watershed Management Villages (NWMV)		
	Water Sources				Water Sources		
	IWMV1	IWMV2	IWMV3	IWMV4	NWMV1	NWMV2	NWMV3
	n= 12	*n*= 12	*n*= 12	*n*= 9	*n*= 12	*n*= 12	*n*= 7
pH	7.2	7.2	7.1	7.2	7.1	7.2	7.3
	(7.0–7.3)	(7.0–7.4)	(7.0–7.3)	(7.0–7.4)	(7.0–7.3)	(7.0–7.4)	(7.0–7.4)
Turbidity [a] (NTU)	2.5	13.7	0 *	0.6	1.3	3.3	1.3
	(0–5)	(5–25)		(0–5)	(0–5)	(0–10)	(0–5)
Chloride content [b] (ppm)	25	22.5	20.8	20*	20.8	20 *	20 *
	(20–60)	(20–40)	(20–30)		(20–30)		
Total hardness[c](ppm)	98	106	79	83.3	54	56	81
	(50–125)	(75–200)	(50–175)	(75–125)	(50–100)	(50–175)	(50–150)
Nitrate content[d](ppm)	15.8	12.3	10*	10*	10*	10*	10*
	(10–45)	(10–45)					

Quality of water has always an impact on human health. In general, lack of infrastructure for safe water facilities, makes tribal people prone to suffer health hazards due to water contaminants, both physico-chemical and microbiological [18]. The World Health Organization (WHO) has recommended levels for water quality parameters that are safe for human health. Two limits have been defined, safe acceptable limit (maximum concentration generally acceptable to consumers) and allowable acceptable limit (maximum concentration beyond which potability is seriously impaired) [19]. During our study, the physico-chemical water quality parameters in both IWMV and NWMV were within safe allowable limits for consumption. While the turbidity level was within acceptable limits (5 NTU) for other wells, it ranged up to 25 NTU in IWMV2. However, this turbidity level was also within allowable limit. IWMV2 was an old well, constructed prior to implementation of the IWMP, needing some repairs, but still used by the people. The turbidity that occurred in rainy season in this particular well might be due to leaching or surface runoff. This finding indicates that as part of the IWMP, attention should be paid to old wells which should be repaired to perfection.

The nitrate content was comparatively higher in IWMV during the rainy months of June and July. The implementation of IWMP results in increased agricultural activity. In the area of our study, the main agricultural activity occurred in the rainy season. During rainy season, leaching of substances occurs from farmland waste in groundwater through percolation and results in increased nitrate content [20,21]. The increased nitrate content in the IWMV wells might be due to the increased agricultural activity in IWMV.

Table 2 shows the month-wise bacterial contamination level in each water source with coliform count/100 mL of water along with antibiotic resistance patters of the detected E. coli. In the case of IWMV, in 17 samples out of 45 (38%), coliform contamination was detected. This figure was 18 out of 31 (58%) in case of NWMV. The contamination level measured as coliform count per 100 mL was in the range of 0 to 7,900 for IWMV and 0 to 13,000 for NWMV. The odds of water samples to have coliform contamination was 2.3 times higher in NWMV as compared to IWMV (95% CI 0.81–6.45, $p = 0.081$). The P value for this difference showed borderline significance and a wide confidence interval indicated the results to be tilted towards IWMV suggesting a contribution of components of IWMP on reduction in bacterial contamination of water.

Table 2. Month-wise coliform count/100 mL water and antibiotic resistance of Escherichia coli isolated from water sources of tribal villages with and without implementation of integrated watershed management programme.

Month	Integrated Watershed Management Villages (IWMV)				Non-Watershed Management Villages (NWMV)		
	Coliform Count/100 mL				Coliform Count/100 mL		
	S/R (Resistance to Mentioned Antibiotics)				S/R (Resistance to Mentioned Antibiotics)		
	Water Sources				Water Sources		
	IWMV1	IWMV2	IWMV3	IWMV4	NWMV1	NWMV2	**NWMV3**
2011/07	170	7900	0	33	140	41	7000
	R (NA, TE)	S		S	R (CEP)	S	**S**
2011/08	14	0	0	0	7.8	7.8	0
	R (NA, TE)				R (NIT)	S	
2011/09	130	22	220	0	0	13	13000
	R (NA)	R (AMP, TI, PI)	S			S	R (AMP, TI, PI64, TE)
2011/10	350	240	0	350	0	79	920
	S	S		S		S	S
2011/11	0	7.8	280	140	0	11	42
		S	R (NA, TE, NIT)	S		S	S
2011/12	0	0	0	0	140	0	130
					S		R (NA, CEP)
2012/01	0	0	0	33	0	140	280
				R (CEP)		S	S

2012/02	350	0	0	140	0	350	a
	S			S		R (AMP, TI, PI, NA, TE, COT, NIT)	
2012/03	0	0	0	0	0	130	a
						R (AMP, TI, PI, NA, TE, COT, AT, All cephalosporins-ESBL)	
2012/04	94	0	0	a	0	0	a
	S						
2012/05	0	0	0	a	130	0	a
					R (CEP)		
2012/06	0	0	0	a	0	0	a
Frequency of contamination (%)	6/12	4/12	2/12	5/9	4/12	8/12	6/7
	(50)	(33)	(17)	(56)	(33)	(67)	(86)

According to the recommendations by the WHO and Central Pollution Control Board of India, drinking water should be free from coliform contamination to be considered safe for use [19,22]. It has been observed that improper sanitation practices are responsible for bacterial contamination of water [2]. The results of coliform count in water from our study (Table 2) show that water from both IWMV and NWMV was contaminated on several occasions. However, the number of times contamination was detected and the count of coliform were lower in IWMV. Awareness about hygiene and improved sanitation practices due to the implementation of IWMP may be a contributory factor for the reduction in bacterial contamination of water in IWMV [23]. The number of households having toilets were significantly higher in IWMV (83/142, 58%) compared to NWMV (27/144, 19%), where open air defecation was the normal practice. In both types of villages, 85% of households had cattle. Animal litter can also be a potential source of water contamination in these villages, which may have resulted in the contamination of water in IWMV as well as in NWMV especially in the rainy season.

The study villages were located on hills. The wells were the only source of water in the villages. Because of the elevated location, the wells normally dried up in the summer season. One of the expected outcomes in villages with "integrated watershed management implementation" is increase in periods of water availability in wells and less dry periods compared to NWMV. It has earlier been found that water sources that become dry in the summer season remain more contaminated throughout the year [24]. Thus, due to integrated watershed management, in the implementation area, dry period for wells as well as contamination of their water would be expected to be low. Of the included water sources, one well out of three in NWMV and one well out of four in IWMV dried up during the dry season. NWMV3 remained dry for five months during the study period and was also found to be the most frequently contaminated water source (86% samples). As compared to this, IWMV3, which was dry for three months, was found to be less contaminated (56% samples). In several areas of the world, as in India, the tribal people reside in hilly terrains and forest areas. There they have to face severe water scarcity during summer season [24], with wells having scanty water reserve and hence contaminated water. Integrated watershed management aims to make water sources available that can supply water throughout the year. This could help to decrease the contamination of water as fewer sources dry up during the dry season.

The proportion of inhabitants treated for diarrhoea at the PHC subcentres was significantly lower ($p = 0.001$) in the IWMV (18/663) compared to the NWMV (42/639). All the diarrhoeal cases in IWMV were in wet season while in NWMV, 23 cases were in wet season and the remaining 19 were in dry season. It is known that tribal people suffer from diarrhoea due to improper sanitation practices and contamination of water [18,25]. Drying of water sources is one of the responsible factors for contamination of water [24]. It has been stated earlier that due to implementation of IWMP, the number of households with toilets were higher in IWMV reducing their open air defecation practice. The wells of IWMV were also dry for lesser periods and less contaminated. Our earlier study in these villages showed that tribal people in the study setting perceived that implementation of IWMP improves the quantity and quality of water in water sources and they also perceived reduction in occurrences of diarrhoea due to it [23].

Total antibiotics delivered to PHC sub-centres in both types of villages are presented as DDDs per 1,000 inhabitants per year in Table 3. The figures were higher in IWMV as compared to NWMV during the study period. This was also true for the previous three years. Recent data on antibiotic dispensing in the community in European countries indicate the figures to be varying from 3,832 to 14,089 DDDs/1,000 inhabitants/year [26]. Population based data, to date, to our knowledge are not reported from India [27]. In our study villages, we could use the existing population as denominator. Antibiotic dispensing was low in our study villages, but between the two settings it was comparatively higher in IWMV.

Table 3. Data on yearly average antibiotics delivered (DDDs/1,000 inhabitants/year) to sub-centres of Primary Health Centre during the years 2008–2011 and 2011–2012 in tribal villages with and without the implementation of integrated watershed

Antibiotic, Assigned DDD	2008–2011 (June–May)		2011–2012 (June–May)	
	IWMV	NWMV	IWMV	NWMV
Tetracycline, 1 g	458	391	302	39
Ampicillin, 2 g	6	--	--	--
Amoxicillin, 1 g	163	33	38	58
Co-trimoxazole, 2 g	267	225	124	99
Erythromycin, 1 g	--	27	--	--
Ciprofloxacin, 1 g	7	52	--	--
Norfloxacin, 0.8 g	75	47	189	156
Nalidixic Acid, 4 g	--	--	1.5	--
Total DDDs/1,000 inhabitants/year	976	775	654	352

Based on our earlier study [23] the reasons for increased antibiotic dispensing in IWMV could be proposed as follows. An important outcome of implementation of IWMP is the socio-economic development that takes place in the target area. This results in increased health awareness and reduced migration. Thus, in the IWMV there is an increased awareness about utilizing health services which might have led to more antibiotic prescriptions than in NWMV. Further, because of less economic development, people from NWMV migrate more frequently for work outside their villages than people from IWMV, thus the population staying throughout the year in IWMV is higher as compared to NWMV. There are also differences in seeking traditional healing measures between the two settings. It was reported that the people from NWMV believed more in using traditional measures compared to people from IWMV [23]. All these factors might have resulted in more delivery of antibiotics in IWMV.

Our findings showed that in the study area 63% (IWMV 65%, NWMV 61%) of E. coli isolates were susceptible to all tested antibiotics (Table 4). A recent study from Nigeria showed that all the 35 E. coli isolated from water samples from open dug wells were resistant to at least one antibiotic [28]. A study from India shows that 96% out of the 151 E. coli isolates from household drinking water (where tube wells and open wells were main sources of water) were resistant to at least one antibiotic [29]. A study conducted on water samples from hilly areas of Uttarakhand State, India, showed that all the 20 E. coli isolates were resistant to at least one antibiotic and 90% of isolates were MDR [30]. A study from Pakistan showed that of the 27 E. coli isolated from drinking water samples 93% were resistant to one antibiotic and 63% of the isolates were MDR [31]. A study from China showed that out of 114 E. coli isolated from surface (lake) water samples, 85% were resistant to one antibiotic [32]. In these five references analyzing susceptibility of E. coli isolates from water samples, the overall susceptibility indicated was 0%, 4%, 10%, 7% and 15% respectively, very low figures compared to the overall 63% susceptibility in our E. coli isolates

Table 4. Co-resistance and Multi Drug Resistance (MDR) of Escherichia coli from well water of tribal villages with and without the implementation of integrated watershed management programme

Resistance	IWMV (*N*= 17)	NWMV (*N*= 18)
	n(%)	*n*(%)
Resistant to one group of antibiotics	3 (18)	3 (17)
Resistant to two groups of antibiotics (co-resistance)	2 (11)	2 (11)
Resistant to three or more than three groups of antibiotics (MDR)	1 (6)	2 (11)
Susceptible to all	11 (65)	11 (61)

The number of samples containing antibiotic resistant E. coli was similar between the two types of villages, with 6/17 in IWMV and 7/18 in NWMV. Amongst the resistant E. coli isolates three isolates from both the settings i.e., IWMV and NWMV were resistant to only one group of antibiotics (Table 4). Two isolates from NWMV were MDR compared to a single MDR isolate in IWMV. In comparison with other studies, MDR was low in the E. coli isolated from the water sources of the study villages [30,31]. There was one isolate from NWMV which produced ESBL (Extended Spectrum Beta-Lactamases) (Table 2).

The isolates listed in Table 5 are representative E. coli isolates from each of the 17 contamination events observed in the IWMV wells and 18 contamination events observed in the NWMV wells. Table 5 shows the antibiotic resistance of these E. coli isolates to antibiotics of different groups. Resistance to tetracycline and nalidixic acid was most common in both the settings. Nalidixic acid resistance is the first step towards the quinolone resistance. High frequency of resistance to tetracycline and quinolone was also recorded in another part of India [29]. The resistance to tetracycline and quinolones has been explained as due to their lower degradation rates (half-life 100 days) in the environment [33]. The longer exposure of bacteria to antibiotic residues in aquatic environment creates a selective pressure and increases the risk of development of resistance [34,35]

Table 5. Antibiotic resistance of Escherichia coli isolated from well water in tribal villages with and without the implementation of integrated watershed management programme

Antibiotics	IWMV (*N*=17) *n*	NWMV (*N*=18) *n*
Tetracycline	3	3
Ampicillin	1	3
Cephalothin (1st)	1	4
Cefuroxime (2nd)	0	1
Cefotaxime (3rd)	0	1
Cefepime (4th)	0	0
Aztreonam	0	1
Meropenam	0	0
Co-trimoxazole	0	2
Amikacin	0	0
Nalidixic acid	4	3
Nitrofurantoin	1	2

Table 5 shows that E. coli isolates from NWMV exhibited resistance to wider range of antibiotics. Resistance was recorded to 2nd and 3rd generation cephalosporins, nitrofurantoin and aztreonam in NWMV, which were not delivered in these villages. Co-trimoxazole, which contains trimethoprim, was the second most delivered antibiotic in these villages. In an earlier study, it was found that there is a cross-correlation between usage of one antibiotic and resistance to another, particularly usage of trimethoprim and resistance to ampicillin and usage of trimethoprim and resistance to cephalosporins [36]. Plasmid mediated resistance and horizontal transfer of resistance between bacteria have also been implicated for transfer of resistance [37,38].

It is difficult to explain the observed resistance to nitrofurantoin and aztreonam, however, the phenomenon of development of cross-resistance is not fully understood. Seasonal migration in tribal areas is also reported to be responsible for many public health issues [23]. If migration has any role in this, in terms of carrying resistant bacteria from migratory sites to the respective native villages, then an integrated watershed management policy which might be helpful in reduction of migration by generation of employment opportunities in tribal areas might thus become also helpful in curtailing the spread of antibiotic resistance. As per our knowledge, this is the first study on contamination of water and occurrence of antibiotic resistance in E. coli from water sources in hilly tribal areas, in the context of watershed management. The strength of this study is that all the water sources in the included tribal villages were followed for a whole year. Secondly, we maintained as much similarity as possible with respect to geographical and social patterns between the villages by selecting villages in a short radial distance. There were also some limitations of this study. As integrated watershed management was the criterion behind selection of villages and it was necessary to choose villages in the same geographical locality for appropriate comparison, there were very few integrated watershed management villages available for the study. That was the reason for the relatively small sample size. In the study villages, all the people generally relied on PHC sub-centres for health services including medicines, so we cannot exclude that people at some occasions might have received medicines from other sources. Seasonal migration for employment by the people, particularly in NWMV may also have affected some parameters of the study as explained above.

CONCLUSIONS

This study found that amongst the included villages, bacterial contamination of water and diarrhoeal cases in the community were lower in villages in which integrated watershed management was implemented compared to villages where it had not been implemented. E. coli isolates from non-watershed management villages showed resistance to a wider range of antibiotics. Collectively, there was a high level of susceptibility (63%) to antibiotics in the analyzed E. coli isolates from the water sources of the hilly tribal villages in study area. This information is valuable in the context of reports of high antibiotic resistance all over the world.

ACKNOWLEDGMENTS

The first author was supported for this work by a scholarship from Erasmus Mundus External Cooperation Window Lot 13 India. We are thankful to BAIF development research foundation and medical officers and employees of the State Health Department of Jawhar Block for their cooperation. Thanks are due to the Medical Director of R. D. Gardi Medical College, Ujjain, V. K. Mahadik, for his encouragement and support during the conduct of these studies. We thank Ashwamedh Laboratory for their contribution to the laboratory work. We also thank the study associates for their help in sample collection.

APPENDIX

Table A1. List of antibiotics used for susceptibility testing using the Vitek 2 AST (antibiotic susceptibility test) cards.

Card No.	AST GN 25	Card No.-	AST EXN7
Sr. No	Antimicrobial	Sr. No	Antimicrobial
1	ESBL	1	Amoxicillin/Clavulanic acid
2	Ampicillin	2	Ticarcillin
3	Ampicillin/Sulbactum	3	Ticarcillin/Clavulanic acid
4	Piperacillin/Tazobactum	4	Piperacillin
5	Cefazolin	5	Cephalothin
6	Cefriaxone	6	Cefuroxime
7	Cefepime	7	Cefuroxime Axetil
8	Aztreonam	8	Cefotetan
10	Imipenam	10	Cefotaxime
11	Meropenem	11	Ceftizoxime
12	Amikacin	12	Aztreonam
13	Gentamicin	13	Meropenem
14	Tobramicin	14	Nalidixic Acid

15	Ciprofloxacin	15	Moxifloxacin
16	Moxifloxacin	16	Norfloxacin
17	Tigecycline	17	Tetracycline
18	Nitrofurantoin	18	Tigecycline
19	Trimethoprim/ Sulfamethoxazole		

REFERENCES

1. Government of India: Ministry of Tribal Affairs . The National Tribal Policy: A Policy for the Scheduled Tribes of India. Government of India, Ministry of Tribal Affairs; New Delhi, India: 2004.
2. World Health Organization Fact Sheet 2.2. [(accessed on 23 August 2013)]. Available online: http://www.who.int/water_sanita-tion_health/hygiene/emergencies/fs2_2.pdf.
3. World Health Organization The Evolving Threat of Antimicrobial Resistance: Options for Action. 2012. [(accessed on 17 January 2014)]. Available online: http://www.who.int/patientsafety/implementation/amr/publication/en/
4. Kalter H.D., Gilman R.H., Moulton M.H., Cullotte A.R., Cabrera L., Velapatino B. Risk factors for antibiotic resistant Escherichia coli carriage in young children in Peru: Community based cross-sectional prevalence study. Am. J. Trop. Med. Hyg. 2010;82:879–888. doi: 10.4269/ajtmh.2010.09-0143.
5. Permanent Solution on Water Scarcity—Watershed Management. [(accessed on 23 April 2014)]. Available online: http://www.old.kerala.gov.in/keralcalljuly04/p17-19.pdf.
6. Common Guidelines for Watershed Development Projects-2008. revised edition 2011. National Rainfed Area Authority, Planning commission, Government of India; New Delhi, India: 2011.
7. Ibekwe M.A., Murinda S.E., Graves A.K. Genetic diversity and antimicrobial resistance of Escherichia coli from human and animal sources uncovers multiple resistances from human sources. PLoS One. 2011;6:e20819. doi: 10.1371/journal.pone.0020819.

8. Introduction to Drinking Water Quality Testing. [(accessed on 23 April 2014)]. Available online: http://www.rdic.org/CAWST-Intro-to-Drinking-Water-Quality-Testing.pdf.
9. Himedia Laboratories. [(accessed on 12 April 2014)]. Available online: http://www.himedialabs.com/HML/images/literature/pdf/100000021/23.pdf.
10. American Public Health Association . Standard Methods for the Examination of Water and Wastewater. 20th ed. American Public Health Association; Washington, DC, USA: 1998.
11. Batty-Smith C.G. The Eijkman test for faecal coli in the bacteriological examination of water supplies. J. Hyg. (Lond.) 1942;42:55–98. doi: 10.1017/S0022172400012638.
12. Vazquez M.P., Antonio O., Saz B.S., Loza E., Baquero F., Canton R. Performance of the VITEK2 system for identification and susceptibility testing of routine Enterobacteriaceae clinical isolates. Int. J. Antimicrobiol. Agents. 2001;17:371–376. doi: 10.1016/S0924-8579(01)00318-1.
13. Carvec S., Cremet L., Reynaud A., Lepelletier D., Caroff N. Vitek2 system: A reliable tool to detect qnr determinants in Enterobacteriaceae without quinolone resistance determining region modifications. Diagn. Microbiol. Infect. Dis. 2009;64:455–457. doi: 10.1016/j.diagmicrobio.2009.03.026.
14. Winstanley T., Courvalin P. Expert systems in clinical microbiology. Clin. Microbiol. Rev. 2011;24:515–556. doi: 10.1128/CMR.00061-10.
15. Clinical and Laboratory Standards Institute (CLSI) Performance Standards for Antimicrobial Susceptibility Testing. CLSI; Wayne, PA, USA: 2009. Nineteenth Information Supplement (Document M100-S19)
16. Magiorakos A.-P., Srinivasan A., Carey R.B., Carmeli Y., Falagas M.E., Giske C.G., Harbarth S., Hindler J.F., Kahlmeter G., Olsson-Liljequist B., et al. Multidrug-resistant, extensively drug resistant and pandrug resistant bacteria: An international expert proposal for interim standard definitions for acquired resistance. Clin. Microbiol. Infect. 2011;18:268–281.

17. Guidelines for ATC Classification with DDD Assignment 2013. WHO Collaborating Centre for Drug Statistics Methodology; Oslo, Norway: 2012.
18. Balgir R.S. Tribal Health Problems, Disease Burden and Ameliorative Challenges in Tribal Communities with Special Emphasis on Tribes of Orissa; Proceedings of the National Symposium on Tribal Health; Jabalpur, India. 19–20 October 2006; Jabalpur, India: Regional Medical Research Centre for Tribals, Indian Council of Medical Research; 2007. pp. 161–176.
19. Cox C.S. Operation and Control of Water Treatment and Processes. World Health Organisation; Geneva, Switzerland: 1964. (World Health Organisation Monographs Series 49).
20. Chen X., Wo F., Chen C., Fang K. Seasonal changes in the concentrations of nitrogen and phosphorus in farmland drainage and groundwater of the Taihu Lake region of China. Environ. Monit. Assess. 2010;169:159–168. doi: 10.1007/s10661-009-1159-3.
21. Nolan B.T., Ruddy B.C., Hitt K.J., Helsel D.R. A national look on nitrate contamination of groundwater. Water Cond. Purif. 1998;39:76–79.
22. Kumar M., Puri A. A review of permissible limits of drinking water. Indian J. Occup. Med. 2012;16:40–44. doi: 10.4103/0019-5278.99696.
23. Nerkar S.S., Tamhankar A.J., Johansson E., Stålsby Lundborg C. Improvement in health and empowerment of families as a result of watershed management in a tribal area of Maharashtra, India. Int. Health Hum. Rights. 2013;13:42. doi: 10.1186/1472-698X-13-42.
24. Hegde N.G. Management of Natural Resources for Sustainable Livelihood- BAIF's Approach. In: Sundaram K.V., Moni M., Jha M.M., editors. Natural Resources Management and Livelihood Security: Survival Strategies and Sustainable Policies. Bhoovigyan Vikas Foundation; New Delhi, India: 2004. pp. 1–17.
25. Kamble S.Y., More T.N., Patil S.R., Pawar S.G., Bindurani R., Bodhankar S.L. Plants used by the tribes of northwest Maharashtra for the treatment of gastrointestinal disorders. Indian J. Tradit. Knowl. 2007;7:321–325.

26. European Centre for Disease Prevention and Control. [(accessed on 12 April 2014)]. Available online: http://ecdc.europa.eu/en/activities/surveillance/ESAC-Net/publications/Documents/ESAC_Yearbook_2009.pdf.

27. Chandy S.J., Thomas K., Mathai E., Antonisamy B., Holloway K., Stålsby Lundborg C. Patterns of antibiotic use in the community and challenges of antibiotic surveillance in a lower-middle-income country setting: A repeated cross-sectional study in Vellore, south India. J. Antimicrob. Chemother. 2013;68:229–236. doi: 10.1093/jac/dks355.

28. Efuntoye M.O., Apnapa M.O. Status of contamination and antibiotic resistance of bacteria from well water in Ago-Iwoye, Nigeria. J. Appl. Biosci. 2010;35:2244–2250.

29. Sahoo K.C., Tamhankar A.J., Sahoo S., Sahu P., Rosales Klintz S., Stålsby Lundborg C. Geographical variation in antibiotic-resistant Escherichia coli isolates from stool, cow-dung and drinking water. Int. J. Environ. Res. Public Health. 2012;9:746–759. doi: 10.3390/ijerph9030746.

30. Chatterjee R., Sinha S., Aggrawal S., Dimri A.G., Singh D., Goyal P., Chauhan A., Aggrawal M.L., Chacko K.M. Studies on susceptibility and resistance patterns of various E. coli isolated from different water samples against clinically significant antibiotics. Int. J. Bioassay. 2012;50:156–161.

31. Patoli A.A., Patoli B.B., Mehraj V. High Prevalence of Multidrug resistant E. coli in drinking water sample from Hyderabad. Gomal J. Med. Sci. 2010;8:23–26.

32. Wang C., Gu X., Zhang S., Wang P., Guo C., Gu J. Charactrization of antibiotic-resistance Escherichia coli isolates from a lake. Arch. Environ. Contam. Toxicol. 2013;65:635–641. doi: 10.1007/s00244-013-9932-2.

33. Seyfried E.E., Newton R.J., Rubert K.F., Pedersen J.A., Mcmahon K.D. Occurrence of tetracycline resistance genes in aquaculture facilities with varying use of oxytetracycline. Microb. Ecol. 2010;59:799–807. doi: 10.1007/s00248-009-9624-7.

34. Wu N., Quio M., Zhang B., Cheng W.D., Zhu Y.G. Abundance and diversity of tetracycline resistance genes in soils adjacent to representative swine feedlots in China. Environ. Sci. Technol. 2010; 44:6933–6939. doi: 10.1021/es1007802.

35. Silbergeld E.K., Graham J., Price L.B. Industrial food animal production, antimicrobial resistance, and human health. Annu. Rev. Public Health. 2008;29:151–169. doi: 10.1146/annurev.publhealth.29.020907.090904.
36. Howard A.J., Magee J.T., Fitzgerald K.A., Dunstan F.D.J. Factors associated with antibiotic resistance in coliform organisms from community urinary tract infection in Wales. J. Antimicrob. Chemother. 2001;47:305–313. doi: 10.1093/jac/47.3.305.
37. Li X.Z. Quinolone resistance in bacteria: Emphasis on plasmid mediated mechanisms. Int. J. Antimicrob. Agents. 2005;25:453–463. doi: 10.1016/j.ijantimicag.2005.04.002.
38. Hopkins K.L., Davies R.H., Threlfall E.J. Mechanism of quinolone resistance in Escherichia coli and Salmonella, recent developments. Int. J. Antimicrob. Agents. 2005;25:358–373. doi: 10.1016/j.ijantimicag.2005.02.006.

Citations

CHAPTER 1

Luo Y, Zhang M (2011) Environmental Modeling and Exposure Assessment of Sediment-Associated Pyrethroids in an Agricultural Watershed. PLoS ONE 6(1): e15794. doi:10.1371/journal.pone.0015794

CHAPTER 2

Chen J, Lu J (2014) Effects of Land Use, Topography and Socio-Economic Factors on River Water Quality in a Mountainous Watershed with Intensive Agricultural Production in East China. PLoS ONE 9(8): e102714. doi:10.1371/journal.pone.0102714

CHAPTER 3

Reyhan Erdogan (2013). Stakeholder Involvement in Sustainable Watershed Management, Advances in Landscape Architecture, Dr. Murat Ozyavuz (Ed.), ISBN: 978-953-51-1167-2, InTech, DOI: 10.5772/55798.

CHAPTER 4

Kovalenko KE, Brady VJ, Ciborowski JJH, Ilyushkin S, Johnson LB (2014) Functional Changes in Littoral Macroinvertebrate Communities in Response to Watershed-Level Anthropogenic Stress. PLoS ONE 9(7): e101499.

CHAPTER 5

Rezende RS, Petrucio MM, Gonçalves JF Jr (2014) The Effects of Spatial Scale on Breakdown of Leaves in a Tropical Watershed. PLoS ONE 9(5): e97072. doi:10.1371/journal.pone.0097072

CHAPTER 6

Burns DA, Woodruff LG, Bradley PM, Cannon WF (2014) Mercury in the Soil of Two Contrasting Watersheds in the Eastern United States. PLoS ONE 9(2): e86855. doi:10.1371/journal.pone.0086855

CHAPTER 7

Dennis Wichelns, Do Estimates of Water Productivity Enhance Understanding of Farm-Level Water Management?, doi:10.3390/w6040778

CHAPTER 8

Tess Russo Katherine Alfredo and Joshua Fisher, Sustainable Water Management in Urban, Agricultural, and Natural Systems, doi:10.3390/w6123934

CHAPTER 9

Sandeep S. Nerkar, Ashok J. Tamhankar, Smita U. Khedkar, and Cecilia Stålsby Lundborg, Quality of Water and Antibiotic Resistance of Escherichia coli from Water Sources of Hilly Tribal Villages with and without Integrated Watershed Management—A One Year Prospective Study, doi: 10.3390/ijerph110606156

Index